What an incredible book! My head is spinning with the histories and lives of all those pioneers. Getting hold of land, planting, buying, selling, fighting, marrying, divorcing, falling out, making up (or not) and battling on through Prohibition, two world wars and now climate change, wildfire and COVID. It's quite a story! A real hymn to the power of perseverance and, sometimes, blind optimism. And, of course, talent. The only people who seem to have entirely lost out are the original inhabitants of this Garden of Eden, the Wappo, whose sad tale is repeated elsewhere in the world, to be sure, but is tragic nonetheless. Mark Gudgel has written a fabulously informative book. I now know more about the subject than I did before. I am amazed that Gudgel managed to fit so much into so relatively short a book.
—Bella Spurrier

The Judgment of Paris was without question the most significant event that elevated Napa Valley on the world stage for wine. Mark Gudgel's The Rise of Napa Valley Wineries *is the story of the pioneers, farmers and winemakers who landed north of San Francisco starting in the mid-1800s and persevered with grit, passion and a vision for over one hundred years to make wine to rival France, and it is as intimately interesting as it is factual. Gudgel is an excellent storyteller and gives the reader firsthand insights into the trials, tribulations and triumphs that made Napa Valley what it is today. To know wine and this storied history of how Napa became world-renowned is to know the Judgment of Paris.*
—Barry Waitte, president, Oakville Winegrowers Association

In The Rise of Napa Valley Wineries, *Mark Gudgel has written a lively, detailed and very readable history of the development of the Napa Valley into a world-famous wine region, pivoting around the famous Judgment of Paris tasting in 1976. This sort of perspective is invaluable for anyone with an interest in wine—and it's remarkable to think of how recent some of these developments are.*
—Dr. Jamie Goode, author of *Wine Science*, 3rd edition

Mark Gudgel has captured the essence of Napa Valley's storied history and eventual renaissance through a game-changing historical event, the Judgment of Paris. His attention to historical detail portrayed among a cadre of interesting characters and excellent storytelling keep the reader engaged from start to finish.
—Laura Larson, editor, *Napa Valley Life Magazine*

T0280198

As the publisher and editor-in-chief of a magazine that covers the extraordinary food and drink producers throughout Napa, Sonoma and Marin Counties in Northern California, I am often asked to recommend my favorite places to visit in our region. All too often, it seems that visitors to Wine Country leave with a sense that it is a place that just sprang forth, fully formed, as it is today, unaware of the deep and rich history of this place. In The Rise of Napa Valley Wineries, Mark Gudgel provides readers with a peek behind the curtain, beyond the faux chateaus and perfectly manicured vineyards, to the real people whose passion and hard work built this storied region that has come to produce some of the most highly acclaimed wines in the world. An entertaining and educational read!

—Gibson Thomas, publisher and editor-in-chief,
Edible Marin & Wine Country

Gudgel constructs a history of Napa Valley to rival the great superhero pantheons of DC and Marvel Comics, one filled with colorful characters, fantastic challenges, thrilling incidents and centuries of growth displaying one of the most exciting terroirs imaginable.

—Chase Magnett, ComicBook.com

The story of the Judgment of Paris and the rise of Napa Valley wineries teaches how, even in the wine world, certain walls need to come tumbling down, walls based on prejudice, snobbery and preconceptions related to the reputation of certain wine areas, especially when it comes to the so-called Old and New Worlds. Mark Gudgel's one-of-a-kind, careful and brilliant historical reconstruction of a pivotal moment in world winemaking is a paean/praise/hymn to trusting in the new, overcoming bias and being open to different cultures, because wine is meant to travel and be shared; wine is inclusion and diversity.

—Laura Donadoni, TheItalianWineGirl.com

Mark Gudgel has done a great job, passionately but honestly reconstructing the events that have shaped the history of Napa Valley. This is a book that can satisfy the curiosity of those who already know the well-known wine-growing area of the "New World" but that can also serve as an instrument of knowledge for Europeans, who are not very erudite about it.

—Francesco Saverio Russo, wine blogger,
wine educator and creator of Archè 2020 by F.S.R.

The Judgment of Paris was the influential gift that Napa Valley wasn't expecting. This somewhat obscure little tasting that no one had heard much about would launch the wines of Napa Valley to a level of notoriety the early winemakers had only dreamed of. It's a fascinating story, and Mark Gudgel provides a thorough dive into the history not only of the Judgment tasting but also of the early settlers and vintners of Napa Valley who would lay the groundwork for one of the world's greatest wine regions. Through his engaging storytelling, Mark brings this colorful history to life, full of interesting characters and unlikely heroes. This is a must-read for any wine lover.

—Joyce Stavert, executive director, Oakville Winegrowers

The Rise of Napa Valley Wineries is a real page-turner. I have read it three times and plan on a fourth. Author Mark Gudgel's passion for this subject reflects in every word. Most importantly, I admire the historical accuracy. Obviously written from the heart, this book will sit on the shelves of my personal library for years to come.

—Kathryn Bazzoli, Sharpsteen Museum, Calistoga

When people ask what put Napa on the map, the Judgment of Paris is the moment that comes to mind. Mark Gudgel's The Rise of Napa Valley Wineries *explores the tasting and more: the years of work that led up to it, the interconnections of stories and the road to greatness for the region of Napa. An exciting read that is hard to put down, made for wine lovers and those who are keen to learn the story from start to finish.*

—Renée Sferrazza, sommelier, wine personality and journalist

Mark Gudgel's The Rise of Napa Valley Wineries *impressively encapsulates the hundreds of years of history leading to the rises and falls but, ultimately, the rise of the Napa Valley and broader Northern California wine industry. He adds new insights to the Judgment of Paris tasting and expands our lens to include influential winemakers and wineries beyond the two winners on that fateful day in 1976. I'm grateful he's included discussion of some of the most critical issues facing the wine industry today and tomorrow: climate action, building resilience to climate change and growing social equity and diversity.*

—Anna Brittain, executive director, Napa Green

Mark Gudgel has written an in-depth look at why and how Napa Valley became, well, Napa Valley. It gives background and context to a wine competition that transformed how the world looked at California wines. This book offers a view of the people, from winemakers to farmers, whose visions have forever changed the world of wine. This isn't just a book for the wine industry; it's a book for anyone who wants to feel inspired to leave their mark on the world.

—Julia Coney, wine journalist and wine consultant
for American Airlines

Mark Gudgel weaves together a thoughtful and engaging history of Napa Valley that will keep you turning page after page. This brilliant historical reconstruction is a testimonial to the region's perseverance and a reminder to never underestimate the underdog—talent will always come out on top, despite preconceptions and prejudices. If you're looking to learn more about Napa Valley and how it got to where it is today, this book is a must-read.

—Paige Comrie, WineWithPaige.com

THE RISE OF
NAPA VALLEY WINERIES

How the Judgment of Paris Put California Wine on the Map

MARK GUDGEL

THE
History
PRESS

Published by American Palate
A Division of The History Press
Charleston, SC
www.historypress.com

Copyright © 2023 by Mark Gudgel
All rights reserved

First published 2023

Manufactured in the United States

ISBN 9781467151856

Library of Congress Control Number: 2022951600

Perhaps Steven did imagine the possibility *of the outcome of the Paris tasting more fully than anyone else associated with the event. Somewhere in the shadowy recesses of the boundary-land between thought and imagination, and the nameless something which asks "what if?", he must have divined that the structure of his tasting created the possibility of the outcome that actually occurred. For, in the past, no one else had provided for such a possibility.*

—Warren Winiarski

To the fairest one

CONTENTS

ACKNOWLEDGEMENTS

Writing this small book was a large undertaking. All told, it synthesizes hundreds of books, five times as many articles, dozens of podcasts, scores of YouTube videos, over one hundred formal interviews, several hundred more informal ones, days spent poring through archives and thumbing through old notebooks and ledgers, thousands of emails, a dozen trips to and from Napa and, alas, countless brilliant wines enjoyed, for the most part with the people who made them—all in the name of scholarly research, you understand. As such, I owe an enormous debt of gratitude to far more people than will fit on this page. Here are some of them, at least.

A special thank-you to George Taber, without whom we would likely not even know that the Judgment of Paris had taken place. Thank you to the wineries who made the wines tasted in Paris that day and to Patricia Gastaud-Gallagher and the late Steven Spurrier for arranging the tasting to begin with. Without these people, there would be no story and no Napa as we know it.

Thank you to the dozens of people—pioneers, winemakers, historians, immigrants and friends—who agreed to be interviewed for this book, who corresponded with me via phone and email and who shared photographs, documents, stories, meals and ideas along the way.

Thank you to the Napa County Historical Society and also to the St. Helena Historical Society and those at the Sharpsteen Museum of Calistoga History for their time, energy and resources and for all they do to preserve the history of the area.

Thank you to Lin Weber, Anna Brittain, Joyce Stavert and Chase Magnett for your helpful edits and critical eye.

Last on the page but first in my heart, thank you to my children, Titus and Zooey, who not only tolerate my absences when I'm away doing research but who also climb into my lap while I'm at home writing and provide me with the breaks and the inspiration I need to undertake a project of this scope. Thank you also to my wife, Sonja, the backbone of our family. I love you.

INTRODUCTION

On the eastern edge of Omaha, Nebraska, where the streets are still paved with ancient, worn cobblestones and, even to this day, horse-drawn carriages comingle with the Hondas, Teslas, Treks and Schwinns that are all but necessary to get around in the River City, there sits the Old Market, a remnant of a time long past.

On the north side of Howard Street, half a block from where it ends, a tall, narrow wooden door leads into "the Passageway." Down a steep flight of stairs, an indoor alleyway some three stories tall, made of brick and draped in freely growing flora, rises up toward the sky. Past Trini's Mexican restaurant and an art gallery, a small patio rests across from a gurgling fountain. The patio belongs to V. Mertz, objectively the best restaurant between Yountville and Chicago, and has for decades. My wife and I had our first date there— and many more since.

V. Mertz has been the launchpad for countless sommeliers who now work all around the world. The restaurant is managed by an advanced sommelier named Matthew Brown, who is as kind as he is intelligent and who, every Monday, hosts a tasting group. Open to anyone, the group has a handful of regulars, mostly industry professionals and somms preparing for their next exam, as well as the nomadic types who participate more erratically. I'm among the latter.

At tasting group, everyone brings a bottle, wines are tasted blind according to the prescribed methodology and the taster comes to a conclusion. The rest of the group then weighs in briefly with their own assessments, with

Matthew going last so as not to influence others with his more learned opinion. Matthew will regularly blind a wine correctly down to the varietal, region, vintage and sometimes even the producer and explain to the group how he came to his conclusions. At tasting group, everyone takes a turn, everyone is welcome and everybody learns. And sometimes, someone messes with the group, as well.

Alongside some of the greatest wines in the world have been tasted Cabernet Sauvignon from Colorado, La Crescent from the Sandhills of Nebraska, Chardonnay from Wisconsin and countless other enological oddities. Frequently, the wines produced in places not often thought of as wine producers will wow the group. A Colorado Cabernet Sauvignon was once almost universally thought to have been a classified left banker from Bordeaux before it came out of its brown paper bag, while the La Crescent from Nebraska was thought by some to be an Alsatian Riesling, by others an Austrian Grüner Veltliner and by all to be quite good.

Fifty years ago, a tasting group like this one would surely have been tasting one thing: French wines. It was almost universally believed back then that the French made the truly excellent wines of the world, and to the others who dared attempt winemaking, well, bless their little hearts for trying. Bless the hearts of the Italians, the Spanish, the Australians and—especially—the Americans, for thinking that they, too, might make wine worth drinking. Let them try. Everybody who was anybody knew that nobody could ever make wine to match the French.

And then, of course, all of that changed overnight when, in 1976, an Englishman in Paris asked some French judges to taste wines from California alongside their French equivalents. That these highly qualified judges unwittingly preferred the American wines to their famous French counterparts was a scandal, and the world owes a great debt of gratitude to George Taber, the only journalist who bothered attending the event. Taber worked for *TIME* magazine and penned the story; the rest, as they say, is history.

Since 1976, the global wine scene has expanded dramatically, and today, it's easy to find Argentine Malbec and Chilean Cabernet Sauvignon, Oregonian Pinot Noir, South African Pinotage, New Zealand Sauvignon Blanc and countless other international wines in almost every place that sells fermented grape juice in the world. Today, in order to pass the required exams and become a sommelier, one must command extensive knowledge of French wine, yes, but also of those wines from every other major wine-producing nation in the world. Thanks to the tasting that Taber dubbed the

Judgment of Paris in 1976, it is now understood that the great wines of the world are made not only in France but also all around the world. And in tasting groups such as the one at V. Mertz in Omaha, all of these wines are studied and respected.

So, to those who turn their noses up at the idea of bubbles made in New Mexico, Merlot from Missouri, Cabernet from New Jersey and all of the other seemingly unlikely ventures in wine that are occurring all around the world, let the stories that follow in this book offer a word of caution: the idea that the Napa Valley could produce world-class wines was virtually unheard of right up until it wasn't. And while the idea that a Cab Sauv from the western slopes of the Rockies might rock your socks off may seem far-fetched to those who've never had one, those of us who taste at V. Mertz know our history, and history would suggest that we aren't so much delusional as we are ahead of the curve.

Mark Gudgel
Omaha, Nebraska
December 1, 2022

Chapter 1

THE GOLDEN APPLE

Said Hera to Paris, "Award the apple to me and I will give you a great kingship."
Said Athene, "Award the golden apple to me and I will make you the wisest of
men." And Aphrodite came to him and whispered, "Paris, dear Paris, let me be
called the fairest and I will make you beautiful, and the fairest woman in the
world will be your wife." Paris looked on Aphrodite and in his eyes she was the
fairest. To her he gave the golden apple and ever afterwards she was his friend.
But Hera and Athene departed from the company in wrath.
—The Iliad

The Napa River spills tranquilly southward, broadening out as it approaches San Pablo Bay, bisecting two prepossessing ranges of densely forested mountains and, at once, creating a lush and sequestered valley that has become a place of legend. The Vaca Range—the eponymous mountains of the Vaca family, from which the name of Vacaville is also contrived—rises on the eastern side and was once the home of the Wintun people. The easternmost mountain range of California's Coastal Ranges, the Vacas separate the Napa Valley from the gently rolling plains of Yolo County in the north and the Suisun Valley near the south, where the trees are somewhat sparse and the burlap-colored grass turns an extraordinarily vibrant shade of green in the wettest years.

To the west of the river, a second range, the Mayacamas, separates the Napa Valley from the Sonoma Valley and thus from Sonoma County and the great Pacific Ocean. The Mayacamas were named for the Wappo village

Left: Wappo woman with infant. *Photo by I.C. Adams, courtesy the Sharpsteen Museum.*

Opposite: Map of Wappo territory, circa 1925, by A.L. Kroeber. *Photo courtesy the Sharpsteen Museum.*

of Maiya'kma, approximately a mile south of Calistoga, where natural hot springs have attracted people of all sorts for centuries. The Wappo were skilled artisans who used natural plant fibers to weave baskets so tightly that they could hold water. They hunted in the densely wooded Mayacamas, which they had done as the valley's sole inhabitants for thousands of years. The Wappo were endlessly devoted to their children, with one known Wappo mantra being "Respect for elders, honor for children." Wappo children played games on the banks of the Napa River, where they also learned to fish, weave baskets and grow into the leaders of the tribe. Tribal chiefs could

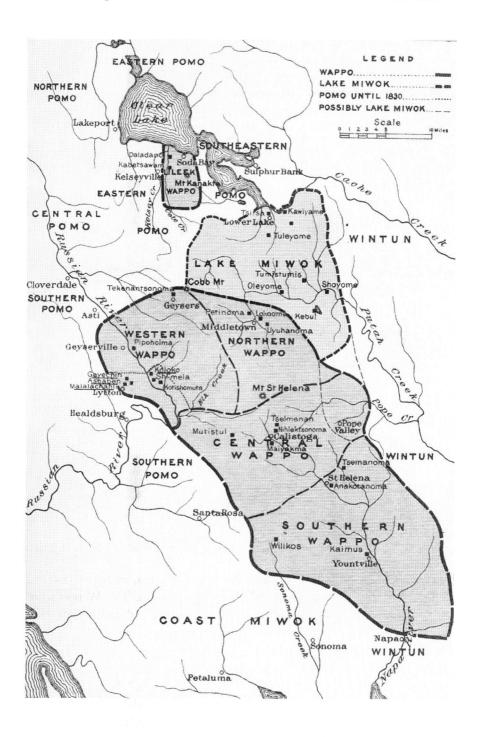

be women among the Wappo, and though they had no written language, their oral histories were rich and beautiful. Thickets of towering redwoods and dense evergreens set deep roots into the rocky soil, binding the earth below the feet of the Wappo tightly to themselves. Today, the sun rises on the Vacas and sets on the Mayacamas, crossing in a perfect arc over the valley floor that is, at its most generously cut, five miles across, just as it has done every day for millions of years.

This valley was once the floor of a sea. During the Mesozoic era, some 140 million years ago, the Farallon Plate collided with the North American Plate, forming a subduction zone and chain of volcanic mountains that would eventually become the Sierra Nevadas. Off the coast, where the Farallon Plate descended into its trench, sediments were scraped off the edges of the plates and accumulated into what would eventually become the Coastal Range. Between these two centers of crustal activity, calmer conditions allowed for the slow accumulation of ocean sediments onto the floor of the future Central Valley. Volcanic activity, the recession of the sea, tectonic uplift, faulting, the organic deposits left behind and, later, great landslides blossoming into alluvial fans all played their part in formulating this tiny geographical appellation that, despite being diminutive and narrow, has the most fertile and diverse soil compositions in the entire world. With the obsidian left over from volcanic activity, the Wappo made points for arrows and spears with which they hunted game. So skilled were the Wappo at crafting these implements that they were traded widely up and down the Pacific coast and even into the mainland.

At the north end of the valley, where the two mountain ranges converge, the peaks of Mount Saint Helena—Mount Mayacamas until it was renamed by a Russian expedition in the early part of the nineteenth century—rise majestically into the sky, high enough to make it one of the few massifs in this balmy region ever to garner any snowfall. It is on the southeastern slope of this formidable landmark that the headwaters of the Napa River can be found. The valley through which the river flows is a mere fifty kilometers long, and that far south of Mount Saint Helena, the San Pablo Bay is home to all manner of birds, fish and mammals. What lies in between these two points is perhaps the closest remaining relation to the Garden of Eden.

That garden, however, this abundantly fertile valley, has been battled over and killed for, inhabited and abandoned, developed and redeveloped, dynamited, set ablaze and fantasized about since the arrival of the Europeans who drove the Wappo from their homes, until at last it became what some always dreamed it might be, what others only feared it

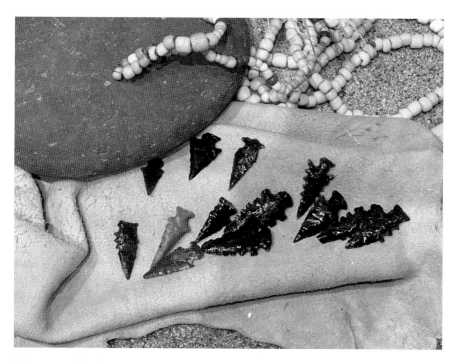

Wappo arrowheads and beads, as displayed in the Sharpsteen Museum in Calistoga. *Photo by author.*

would become, at once sublimely immaculate to some while grotesquely unrecognizable to others and, undeniably, the epicenter of the American viticultural industry. This is the Napa Valley, from whence in 1976 twelve wines from little-known producers went head-to-head with some of the very best of France and, against incredible odds, proved once and for all that great wine can be produced around the globe and not merely in the storied soils of Bourgogne and Bordeaux. That victory, however, was long in the making and came at no small expense, and a great many events, every one another brick in the foundation of this marvelous epic, was fated to occur before the proud and towering walls of Troy, once believed impenetrable by all, would finally be breached.

The Wappo were originally known as the Onasatis and had inhabited the Napa Valley since before King Solomon completed construction of the First Temple in Jerusalem. The Onasatis received this new version of their name from the Spanish *guapo*, meaning "handsome" and also "brave," and they lived in thatched homes made of grasses, typically along the water. The Napa River teemed with fish, from sturgeon to Chinook salmon, rainbow trout, steelhead, perch and more, all of which were easy to catch in the delicate folds and camouflaged lagoons of the sedate and equanimous water. All sorts of birds and mammals could be found around the river and the streams that feed it, many of them feasting on the ubiquitous fish, while the hillsides abounded with deer, fox, elk and other game. In the Wappo tongue, the word for which the western mountain range is named is said to mean "cry of the mountain lion," and the fearsome creatures prowled the shadows in search of prey. Great bears and panthers, too, crept stealthily about the hills; the Wappo shared a mutual respect with these mighty predators, and so man and beast each avoided the keenly sharpened and skillfully wielded points of the other whenever possible. There was more than enough game in this petite yet proliferous valley to supply both man and beast, after all.

It was to this place of prelapsarian beauty, a luscious abundance of great natural scenery, verdant hillsides, dense forests and bountiful resources, that the Spanish arrived early in the nineteenth century. Colonizers from Spain had been influential up and down the Pacific coast of North America for some time prior to their arrival in the Napa Valley. The lower reaches of Mexico were, by then, thoroughly populated, but the farther north one traveled along the Pacific coast, the less influence the Spanish wielded, and by the time one got anywhere near the ranges that encircled the Napa Valley, the influence of Europeans had long since dwindled. Thus, during the sixteenth and seventeenth centuries, while the East Coast of North America bustled with Europeans from New Amsterdam to St. Augustine, the Wappo continued their peaceful and unadulterated existence, until the Spanish eventually made their way north.

Father Junípero Serra is believed by many to have planted California's first vineyard. Father Serra planted Mission grapes in his vineyards, a species of *Vitis vinifera*—European grapes—that had been introduced to Mexico by the Spaniards in the sixteenth century. Over the years, Father Serra established numerous other missions, simultaneously planting vineyards, making wine and ultimately having the moniker Father of California Wine bestowed on him. Thanks to the influence of the Spanish, Roman Catholicism thrived, and missions continued to appear up and down North America's left coast for generations to come, spreading Christianity, subjugating Natives and staking claims from which to fly their flag. But the Spanish soon found that stealing the land from peaceful and comparatively poorly armed Natives was one thing and maintaining control of it was another.

The commercial sale of wine in Mexican territory—including Alta California, an expansive province that included modern-day California, Nevada, Utah and significant sections of four other states to the east—was established by the late eighteenth century. By the early nineteenth century, the industry was regulated by the Mexican authorities, and prices were fixed by the government. As time went by, more and more Mexican territory with a desirable Mediterranean climate was planted to various species of vinifera.

Mariano Guadalupe Vallejo was a stout lad who possessed a look of quiet confidence even in his youth and, soon after, a penchant for growing his sideburns into dense, untamed muttonchops. Vallejo had a round face, a bulbous nose and a privileged upbringing. The eighth of thirteen children, he was born a Spanish subject in Monterrey, which at the beginning of the nineteenth century was located in Alta California. Vallejo was serving as a personal assistant to Governor Luis Argüello when the Mexican War of Independence left the Mexicans free of Spanish rule and struggling to hold their sprawling empire together.

Still a boy, Vallejo joined the Mexican military and rose quickly through the ranks; at the age of twenty-six, he became the military commandant of the San Francisco Presidio. A year later, Vallejo established a hacienda that would become the town of Sonoma, forty-some miles north of San Francisco on the Pacific side of the Mayacamas, assisting the Mexican government in staking a northerly claim to land that would prove difficult to defend.

Vallejo continued to steadily climb through the ranks in the Mexican military, helping his government oversee part of a territory that, after thousands of years of unfaltering tranquility, was fast becoming contested and uncertain. California changed hands repeatedly, first from the Indigenous Natives to the Spaniards and then from the Spaniards to the Mexicans, yet its northern territory remained largely unknown to those who flew their flag above it for much of the early part of the nineteenth century.

The first recorded exploration—or rather, invasion—of the Napa Valley by non-Natives came in 1823 and was led by Don Francis Castro and Padre José Altimira under armed escort, seeking out a location to install yet another mission. Within a decade, George Yount, a pioneering Americano and veteran of the War of 1812, found his way to roughly the same location. Yount was born in North Carolina in 1794 and moved to Missouri in his boyhood. Having failed in his pursuits, Yount left his impoverished wife and their three children and struck out west with fellow trapper William Wolfskill in search of greater prosperity—or at least a fresh start.

Yount was resourceful and at least somewhat skilled in the art of carpentry; somewhere along the line, he learned to make wooden shingles, and in 1834, he arrived in the fledgling community of Sonoma, where he found himself in the personal employment of the town's founder. That same year, the Mexican government that Mariano Guadalupe Vallejo served made the bold move to secularize all missions within its sphere of influence, removing the Natives and the land from the control of the Franciscan missionaries. In the decade prior, the Mexican government had issued fifty large land grants. In the decade that followed, ten times that number were endowed on those citizens who sought to inhabit the distant reaches of the Mexican state. Though the Natives were no longer the victims of missionaries' proselytization and oppression, they benefited little if at all from the secularization of the missions; the vast majority of the grantees of these large parcels of earth were not those indigenous to the land.

In 1836, the same year the Potowatomi, Ottowa and Chippewa were herded off of their land in the United States, now-general Vallejo forced the Wappo to sign a treaty, vacating their rights to the Mayacamas in which they had lived for thousands of years. The opportunistic Yount demonstrated no scruples about taking advantage of the firsthand knowledge of this event passed along to him from his boss and succeeded in convincing General Vallejo to intercede on his behalf. Vallejo persuaded Governor Nicolás Gutiérrez to grant land to the Americano Yount, described by one historian as "brave, illiterate, ugly, and industrious," on the condition that he become a Mexican citizen and a Roman Catholic. In the same year that the Wappo lost their land to the Mexican government, the ugly yet industrious Yount succeeded in bargaining away his religion and citizenship in exchange for 11,887 acres, which he named Rancho Caymus after a subgroup of the Wappo whose land he had acquired.

A second claim was established the following year, rewarding service to the government with enormous tracts of land. Nicolas Higuera, a former soldier, was granted two parcels in the southern end of the valley, significantly past Yount's claim, which he named Rancho Entre Napa and Rancho Rincon de Carneros. Higuera ran thousands of head of cattle on his ranchos, erected some corrals and built a small house.

With the Mayacamas separating Yount's Rancho Caymus from General Vallejo in Sonoma, the former constructed what may have been the first

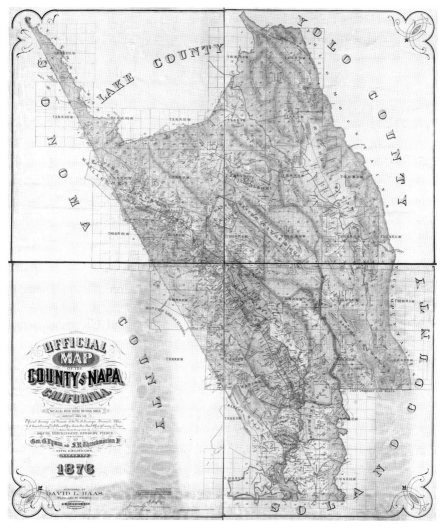

Map of Mexican land grants, known as ranchos, located in the Napa Valley, 1876. *Courtesy Library of Congress.*

wooden house in California and began working the land and running cattle like Higuera. Within a few years, in 1838, Salvador Vallejo, the general's younger brother, was granted twenty-two thousand acres, which he named Rancho Napa and enlarged the following year with the addition of a second grant known as Salvador's Ranch. The days of the Mexican government's passive rule from afar were drawing to a close as private citizens came flooding onto the valley floor, eager to acquire a piece of paradise to call their own.

THE MEXICAN GOVERNMENT, STILL struggling to control its massive territorial holdings, sought to secure its grip on the land by populating it in much the same way that the United States government would seek to populate its way west with the Homestead Act a few decades later. By the early 1840s, ranchos were popping up all over Alta California. The spread-thin and fledgling Mexican government was in near-constant turmoil. For the decade or so leading up to the Bear Flag Revolt and, ultimately, Californian statehood, it was the practice of numerous Mexican governors to parcel out Alta California in measurements of square leagues. A league is equivalent to three miles, and when divvied up in chunks of multiple square leagues to those who found favor with the governor du jour, it didn't take long for the Napa Valley, which measures only around ten leagues north to south and as little as one from east to west, to become the private property of enterprising and often unscrupulous would-be land barons.

In the years to come, a revolving door of governors would bestow one land grant after another. An Englishman named Edward Turner Bale staked his claim on the Mayacamas side of the valley and named his rancho Carne Humana, Spanish for "human flesh." Bale would also establish an impressive, towering mill on his claim, which in the summer of 1846 played host to the men who would stage the Bear Flag Revolt. South and east of Bale's claim, Rancho Yajome was granted to a soldier who had served under General Vallejo by the name of Damaso Rodriguez. Rancho Tulucay was located just south of Rodriguez's Rancho Yajome and was issued to a soldier named Cayetano Juárez who had, the year prior, erected a small adobe on the property.

Edward Turner Bale spent far more time in court than he probably would have preferred and was frequently tied up in lawsuits over everything from malpractice to breach of contract. His legal issues, however, pale in comparison to the relationship he had with the prominent family into which he married. At one point, Bale accused Salvador Vallejo of having an affair with María, Bale's wife and Salvador's niece, and challenged him to a swordfight. Upon being soundly defeated, Bale later attempted to assassinate Vallejo but missed and instead succeeded in shooting Cayetano Juárez, the proprietor of Rancho Tulucay, who miraculously survived the bullet. Surprisingly tolerant of Bale's violent behavior, Vallejo had Bale publicly whipped for the attempt on his life.

Rancho Huichica was nearly parallel to Rancho Tulucay, located on the western side of the valley in the shadow of the Mayacamas. It was issued to Jacob P. Leese, who, like Bale, had married into the Vallejo family. Another grant issued around that time, Rancho Locoallomi, was located as far from Huichica as possible, in the northeastern part of the valley, in the Vaca Range. Rancho Locoallomi was issued to a trapper from Kentucky named William Pope, who named his claim from a Miwok word meaning "place of the cottonwood." The nearby valley became Pope Valley. In the year 1841 alone, five new ranchos totaling more than forty thousand acres were occupied by settlers in the tiny Napa Valley.

Two years later, in 1843, George Yount managed to secure a second, smaller land grant in the Vaca Range, which he named Rancho La Jota. At the foot of the mountain, on Yount's second claim, the Howells settled. John Howell would eventually open a blacksmith shop in Napa County, in nearby St. Helena, and a mountain that was part of Yount's Rancho La Jota would come to bear his name.

North of Bale's Rancho Carne Humana, José de los Santos Berryessa established a rancho that he named Mallacomes. Berryessa's Rancho Mallacomes—also known as Mallacomes y Plan de Aguacaliente, for the nearby hot springs—extended north into Sonoma Country and included the Mallacomes Valley. In 1853, Berryessa sold a significant parcel of his land to a Bear Flagger named Thomas Knight. This parcel included the Mallacomes Valley, north of Napa in Sonoma County. That valley would eventually come to bear Knight's name.

Two other Berryessa brothers, José de Jesús and Sexto, claimed Rancho Las Putas on the far eastern side of the valley that same year. Their claim, named for the Putah Creek that bisected the property, was slightly south of Yount's significantly smaller Rancho La Jota and more or less parallel to Bale's Carne Humana on the opposite side of the valley. In 1843, General Vallejo applied for and was granted the rights to Rancho Suscol, which sprawled to eighty-four thousand acres and overlapped significantly into Solano Country.

Shortly after, Joseph Ballinger Chiles, an army colonel from Kentucky, founded Rancho Catacula east of St. Helena and west of Rancho Las Putas and erected a flour mill in the area. Chiles packaged flour from the mill with the help of John Conn, who had tried but failed to establish his own rancho. One of Chiles's daughters, Fanny, married a man named Jerome C. Davis, and together, Jerome, Fanny and Joseph Chiles established a township, which they called Davisville.

By 1846, the United States and Mexico were on the brink of war; however, that did not prevent Pío Pico, the last Mexican governor of Alta California, from granting José Ygnacio Berryessa the last Mexican land grant in the

Napa Valley, Rancho Chimiles, not far northeast of the miniscule hamlet of Napa. The war stemmed from the United States government's desire to expand the country's borders. President James Polk's personal position deviated substantially from the political opinions that had prevented the United States from annexing Texas when it had gained independence from Mexico a decade earlier, and the eleventh president of the still-fledgling United States was enthusiastic about expanding American territory anywhere he could. Texas was interesting. Oregon, too. But Alta California was in a league of its own. First, Polk offered to purchase it from the Mexican government, and when his offer was refused, he moved troops into disputed border territory, intentionally provoking the Mexicans into a war that Polk was confident the United States would win.

On May 13, at Polk's urging, the United States Congress declared war on Mexico, and Mexico's refusal to reciprocate with a like declaration did nothing to deter U.S. troops from invading. The war lasted until February 1848, when the Treaty of Guadalupe Hidalgo recognized the United States' annexation of Texas and guaranteed sale of Mexico's northern territories, including Alta California, to the United States for $15 million. The Californian cornucopia was moving closer to becoming a part of the United States of America.

Etienne Theé was a French forty-niner who planted a vineyard on the banks of the Guadalupe Creek, in the foothills of the Santa Cruz Mountains, just south of San Jose. It was 1852, and California was now a state—and a popular one, at that. Men like Theé flooded in from around the world, hoping to strike it rich in gold. Most failed to do so, and many pursued other avenues of earning. For his part, Theé began making wine, and his vineyard became one of if not the first commercial winery in the state of California. The Frenchman called his winery Almaden Vineyards. The winery produced French varietals from Theé's homeland and helped slake the thirst of the miners pouring into San Francisco during the gold rush that, in less than two years, saw the Bay Area's population increase twenty-five-fold.

Back in Theé's native France, Emperor Napoleon III was busy implementing a ranking classification, still known as the 1855 Bordeaux Classification, that would impact the wine trade for centuries to come. The emperor tasked the Bordeaux chamber of commerce with ranking the great wines of the region. Though property has changed hands and borders countless times over the years, the 1855 classification remains a leading determinant of price. Thus, while Etienne Theé was almost haphazardly giving birth to the commercial wine industry in California, the commercial wine industry in France was busily placing yet another stone in the wall that guarded the lore of its superiority.

All the same, the terroir of Santa Clara County combined a coastal climate, fertile soil, limited diurnal range and moderate continentality to make it an excellent location for growing grapevines, yet Almaden Vineyards alone could not provide enough fermented grape juice to satisfy the onslaught of miners and pioneers. As evidence arose that those in America's newest state had a thirst for wine and that there was money to be made, the commercial wine industry in California began to expand.

Martha Hunter was born on a plantation in North Carolina but married an army surgeon, Charles Hitchcock, and left behind her privileged life to move with him to New York, where they had a daughter, Lillie. In 1851, the family was transferred to San Francisco and easily blended into the elite, educated socialite class of the excitingly cosmopolitan young port city.

Few parents would feign surprise at the fact that Martha's daughter, Lillie, did not demonstrate the level of concern for what might be seen as "ladylike" that her mother was committed to bearing out in life. One seemingly understated account relays that Lillie "defied the constraints of her conservative upbringing and embodied the spirit of frontier California," while others deal more directly in anecdotes about Lillie's lifelong affinity for cigars, whiskey and gambling. Saved from a fire by San Francisco's Knickerbocker 5 fire brigade, Lillie was made an honorary member and developed a passion for firefighters that would last a lifetime, even embellishing her signature with the number 5. It was evident early on that Lillie Hitchcock, a pretty girl with fierce eyes and a cunning smile, was a force to be reckoned with. Having no desire to reckon with their daughter, Lillie's parents purchased land in the northern end of the Napa Valley and banished her to it. She named her new surroundings Larkmead.

George Yount had planted vineyards in the Napa Valley, and it follows that he was among the first in the valley to make wine. Though it is unlikely that he ever produced any of it for commercial purposes, the Missourian-turned-Mexican-turned-Californian apparently had some level of ability when it came to the art of winemaking. On February 2, 1854, a San Francisco–based newspaper, *Alta California*, wrote of Yount's wine:

We are indebted to Mr. Geo. C. Yount for some wine from his ranch in the upper portion of Napa Valley. The wine is evidently new. It is clear, bright red, and bears a good deal of resemblance to the Bordeaux wines, but is better than most of the claret offered in the San Francisco market, and probably, with more age, and, perhaps, a little better management, would equal the best of French wines.

Whether or not the author of this early review—who held an equal affinity for both wine and commas—was at all qualified to judge quality or not, the concluding statement is yet one more exclamation point in the history of the Napa Valley. The speculation that California could produce wines on par with those of France began early and was oft-repeated, even if such claims were not taken seriously by the most influential forces in the market for generations to come.

North of San Francisco, in 1857, John Patchett became the beneficiary of still more of Higuera's Rancho Entre Napa land grant. Patchett grew grapes on the property and could have repurposed Higuera's old cottage into a makeshift winery. This may have been Napa County's first commercial winery, producing six barrels of wine, which were sold to San Francisco restaurants. The very idea of restaurants was relatively new and somewhat fashionable, as fine dining had arisen only at the end of the French Revolution (1789–99), when the best chefs of France, their employers mostly decapitated, pivoted and began cooking for the masses. That such establishments in San Francisco were carrying wine half a century later not only suggests the influence of French immigrants but also foreshadows the incredible culinary scene that would eventually come to be one of the defining characteristics of the Bay Area.

A year after Patchett opened his winery, he hired a Prussian by the name of Charles Krug, a bespectacled and sublimely clever young man of twenty, to serve as winemaker. Krug, endowed with the pragmatic and efficacious nature of his fellow countrymen, used a cider press he bought off a Hungarian to produce another two hundred gallons the following year.

The Hungarian from whom Krug acquired his apple press was Agoston Haraszthy, an eccentric son of landed gentry from the city of Pest, the comparatively flat sister of the precipitous city of Buda on the Danube River. Haraszthy settled in Sonoma County the same year that Patchett acquired his portion of Rancho Entre Napa. Haraszthy, who went by the self-styled moniker Count of Buena Vista, opened Buena Vista winery near the town of Sonoma. The impressive stone structure was among the first wineries built in California.

Left: Lillie Hitchcock Coit dons a fireman's uniform. *Courtesy Napa County Historical Society.*

Right: Charles Krug. *Courtesy St. Helena Historical Society.*

The same year that Patchett hired Krug, Dr. George Belden Crane established himself near what would become thriving St. Helena. Crane planted a vineyard that would stand the test of time, still producing amazing wines to this day, and built a stately home overlooking it.

Not long after, the Prussian immigrant who cut his teeth working for Patchett and later served as Haraszthy's assistant winemaker at Buena Vista would establish his own winery. The ambitious and intelligent Krug, having gained experience, founded the Charles Krug Winery in 1861. Krug married one of Edward Bale's daughters, Carolina, who provided the Prussian with 540 acres of Rancho Carne Humana as part of her dowry. Patchett could not have been glad for the competition from his talented former employee; he replaced Krug with a Swiss watchmaker-turned-winemaker by the name of Henry Pellett.

The following year, Jacob Schram built his winery farther up the valley floor, nestled on the western side in thick redwoods on hilly slopes. Like Krug, Schram carved away a section of Rancho Carne Humana, just north of Bale's impressive mill. The heavily bearded Schram was a barber by trade and an immigrant from Germany who fancied the remote location of Diamond Mountain. Schram kept his beard trimmed neatly at the sides, which gave his face a hawkish, angular appearance and complemented his intense and penetrating eyes. Schram and his wife, Annie Weaver, worked indefatigably to clear the thickly timbered property in the foothills of the Mayacamas before planting the land to grapes.

By 1876, Schram was producing twelve thousand gallons of wine annually, and it was highly regarded enough to get the attention of Robert Louis Stevenson, who visited the winery in 1880 while honeymooning with his new bride, Fanny. Stevenson, a Scot from Edinburgh born at the advent of Californian statehood, wore his jawline clean-shaven so as to leave his cheekbones visible, revealing a roguish smile. What his jaw lacked in hair, his upper lip made up for, and the thickly mustachioed young man bore some resemblance to his gunslinger contemporary Wild Bill Hickok.

Most of Stevenson's noteworthy works were still in the young writer's head at this stage, and his manuscript about their time in California, *The Silverado Squatters*, was not released until 1884, one year after *Treasure Island*. Stevenson and his new bride enjoyed Schram's wine and hospitality. "The stout, smiling Mrs. Schram…entertained Fanny in the verandah while I was tasting wines in the cellar," wrote Stevenson. By Stevenson's account, Jacob Schram led the tasting and observed his guest's reactions with great interest. "I tasted all," wrote Stevenson, "every variety and shade of Schramberger,"

Unlike her grandfather, Elizabeth Yount was refined, with a dignified appearance, and while she may have been industrious, she was certainly not ugly. Elizabeth had made her way to California to join the family, her aunts and grandparents and cousins, and there she met the ruggedly good-looking Thomas Lewis Rutherford, whom she married. Her father dead and Elizabeth without a dowry, her grandfather gifted the couple one thousand acres at the northern end of Rancho Caymus as a wedding present, and Thomas planted vineyards and began to work the land.

Under Yount's supervision, a town known as Sebastopol had been laid out on Rancho Caymus. Yount may or may not have realized that his township shared its name with another, which had previously been established just over the Mayacamas toward the Pacific. Yount died in 1865, and soon after—in 1867—Sebastopol in the Napa Valley was rechristened Yountville, with the dual purpose of honoring its founder and not confusing those who were attempting to locate the town of Sebastopol in neighboring Sonoma County.

Tiburcio Parrott would one day be known as the outlandishly eccentric landowner who had his coachman blow a French horn as he rode through St. Helena on his way to the brothel, but that day was a long way off.

Robert Louis Stevenson. *Courtesy Napa County Historical Society.*

Tiburcio Parrot was born in Mazatlan, where his father, John, was the U.S. consul. Tiburcio was well educated and well traveled, stylish and well kempt, but his father felt he had a "restless disposition" and called him back to San Francisco to rejoin the family in the early 1860s. In 1868, John Parrott lent money to Agoston Haraszthy to help fund Buena Vista but later used his influence to force the "count" to resign his position, claiming he was inept.

In 1873, Tiburcio and his father became controlling owners of the Sulfur Bank Mine in Lake County, north of the Napa Valley. In these mines, they employed Chinese laborers to dig caves, a common practice widely used in the blossoming wine country. In 1880, the California State Legislature passed a law that forbid the employment of Chinese nationals. Tiburcio made history by openly refusing to abide by the law. He was arrested, imprisoned and tried. In challenging this racist law, Tiburcio Parrott won a victory in the courts, was released from jail and rehired all the Chinese workers who had been laid off during his incarceration.

Jacob and Fritz Beringer terraced seventy acres of vineyard and planted it mostly to vinifera, including Cabernet Sauvignon, Pinot Noir, Zinfandel and Mourvedre. Jacob Beringer had arrived at Ellis Island from Mainz, Germany, in 1868, and quickly made his way west. By 1869, he was in Napa and working for Charles Krug. John Parrott died in 1884, and the following year his son Tiburcio began construction on an immense house. Architect Albert Schropfer had just completed construction of the Rhine House for the Beringer brothers, neighbors whom Tiburcio had nicknamed Los Hermanos, and Schropfer agreed to take on Parrott's equally ambitious project. Under his supervision, the stunning Chateau Miravalle—meaning "valley view"—was erected. A patron of the arts, Tiburcio spared no expense in building his home, and his grand estate and lush grounds rivaled the finest in the valley.

Tiburcio appointed his house with original works of art, including one commissioned from French painter Jules Tavernier in which he had himself and the Baron Rothschild painted into a Pomo sweat ceremony. Parrott was too outlandish for his own good. One evening, while entertaining Los Hermanos at Miravalle, the braggadocious Parrott bet the brothers that his caves would withstand a blast of dynamite. They didn't.

Tiburcio Parrott 1840-1894
Courtesy of Ramona Beringer

Left: Tiburcio Parrott. *Courtesy St. Helena Historical Society.*

Below: Tiburcio Parrott's historic home at the foot of Spring Mountain, Miravalle. *Photo by author.*

Henry Walker Crabb was born in Ohio, but it was in California that he would become famous for his immense collection of grapevines and the vineyards in which he planted them. H.W. Crabb moved to California and, in 1868, bought land near Oakville in the Napa Valley. Crabb built a railroad depot and planted Hermosa Vineyard, producing mostly table grapes. In 1872, Crabb planted a vineyard to wine grapes on a flat expanse of the valley floor and named it To Kalon, a Greek phrase that translates roughly to "the highest beauty." Crabb expanded the vineyard until it covered more than one thousand acres, contained four hundred varietals and produced fifty thousand gallons of wine per year. The wines produced at To Kalon were highly regarded. According to one writer, Crabb was "without peer" as a winemaker in the state of California.

Gustave Niebaum's beard would have given that of Karl Marx a run for its money. Born in Helsinki, Finland, Niebaum often identified as Jewish, though some historians question whether or not he actually was. A ship captain, fur trader and part owner of the Alaska Commercial Company, Niebaum was instrumental in mapping the Alaskan coastline before settling in the Napa Valley, where he hired Hamden McIntyre to design his winery. McIntyre, who had worked as a piano maker and fought in the Civil War, was lured to San Francisco by the promise of employment. A friend of Niebaum, he had no interest in building the mundane and instead set to work erecting a massive winery from stone quarried nearby, with arched doorways, gabled windows and a flair previously reserved for living spaces. Soon, Niebaum's Inglenook was one of the most impressive wineries in the New World— and possibly in the Old World as well. One year later, in 1882, German immigrants named George and Catherine Schonewald purchased land in St. Helena. They built a summer home, which they named Esmeralda, and planted a vineyard of seventeen acres. A few years later, the Schonewalds sold part of their property to Franz Kraft, who built a Victorian-style family home and a sandstone wine cellar. Kraft called his cellar the Old Kraft Winery and retailed his wine to the residents of St. Helena.

In the foothills of the Mayacamas, just west of Crabb's To Kalon, John Benson planted his own parcel of eighty-four acres, enough to produce fifteen thousand gallons of wine, and named his winery Far Niente. Riffing on the Italian phrase *dolce far niente*, meaning "it is sweet to do nothing," he declared his own Far Niente to mean "without a care." Benson was a forty-niner with enough money to hire Hamden McIntyre to build him a grand structure overlooking the valley floor. Benson's winery went up in the base of the Mayacamas, just above To Kalon and other notable properties of the day.

Top: The planting of the To Kalon Vineyard. *Courtesy Graeme MacDonald.*

Bottom: Sketch of the To Kalon Vineyard and H.W. Crabb for the *Register*, circa 1890.
Courtesy Graeme MacDonald.

Not long after completing Far Niente, Hamden McIntyre undertook his most ambitious project, just north of St. Helena on the western side of the valley floor. The 117,000-square-foot, castle-esque, three-story Bourn & Wise Wine Cellar was completed in 1889. Large enough to hold over three million gallons of wine, the immense structure was McIntyre's masterpiece, built into the terraced hillside behind it with striking archways and a tower looming like a sentinel over the valley floor below. Shortly after the Bourn & Wise Wine Cellar was completed, Wise fell ill and sold his shares to Bourn, who made short work of renaming the building Greystone after the gray stones in the front of the building. The largest wine cellar in North America,

Greystone was a wine cooperative where local winemakers could rent space to age their wines.

Everyone needed rope, and Alfred Tubbs had managed to all but corner the market in San Francisco. Gold miners, maritime workers and countless others were in need of the stuff, and Tubbs was happy to supply them. After making his fortune, the austere, bearded, balding Tubbs toured Europe in retirement. In France, he fell in love with wine, and on his return to California, he entered into the industry himself.

Tubbs purchased 254 acres near Calistoga in 1882 and erected a wooden winery, which he dubbed Hillcrest Estate. He soon began production under the lingering gaze of Mount Saint Helena. Unfortunately, Tubbs's winery burned down four years later. Undeterred, Tubbs built a new winery of less flammable material, a stone chateau with walls twelve feet thick in places. This time, vanity got the better of him, and he named it the A.L. Tubbs Winery. In the decades to come, Tubbs's winery would grow to be one of the most productive facilities in California.

Down valley, Seneca Ewer was a state senator who helped form the Bank of St. Helena and became one of eleven men to sit on its board. The bank was founded using capital derived almost exclusively from winemaking, and Ewer was juxtaposed on the board with such notable vintners as Krug and Niebaum. Ewer himself was in partnership with the Atkinson family, and together, they raised yet another stone monument to viticulture, the Ewer and Atkinson Winery, in the area north of Yountville by then known as Rutherford.

All across the Napa Valley—and the entire state of California—land was being gobbled up, and much of it was being planted with grapes and other kinds of fruit. On Spring Mountain, high above the location where Tiburcio Parrott and the Beringer brothers had established their wineries, George Cook acquired the deed to a small plot of land, thick with redwoods and madrones and with a spectacular view of Mount St. Helena and the valley floor below. Cook, too, planted grapes.

Down the mountain from Cook's flat, Josephine Tychson and her husband, John, had purchased 147 acres in 1881 but hadn't done much with it. John died by suicide in 1886, leaving Josephine widowed. The resourceful and intelligent Josephine installed a redwood winery and, in so doing, likely became the first woman winemaker in California. Tychson had a love of equestrianism to rival her appreciation for viticulture and divided her time between riding and overseeing production at Tychson Cellars.

Certificate deeding land on Spring Mountain to George Cook by President Chester Arthur, dated December 5, 1884. *Courtesy Smith-Madrone.*

Farther down valley from Tychson's new construction, George and James Goodman commissioned a three-story gravity-fed winery across the street from the Oak Knoll train station. The Goodman brothers also hired architect Hamden McIntyre to build their winery. McIntyre's Inglenook, Far Niente and Greystone wineries had clearly caught their attention, but the brothers presented McIntyre with a new challenge: build another awe-striking monument to viticulture, but instead of the traditional quarried stone that McIntyre was used to harvesting from the nearby hillsides, build this one instead out of native trees—Douglas fir and redwood. McIntyre had previously built gravity-fed wineries into hillsides, a pragmatic approach that offered those with tremendously heavy loads of grapes relative ease of access to the top floor, but the Goodmans' claim was in a grove of oak trees on the Eschol Ranch, flat land that lacked a hillside to build into, presenting further challenges.

McIntyre succeeded in building a state-of-the-art winery for the Goodman brothers, yet another of his architectural wonders, this one made of wood and fashioned complete with elevators drawn by horses for raising grapes to the third floor in place of a hillside build-in. The Eschol Winery was born. Soon after, in 1888, the Goodmans entered their Cabernet Sauvignon and Sauterne—a sweet, botrytised wine in the Bordeaux style—in the California State Viticultural Convention, where both wines took first prize. The following year—at the Paris Exhibition, where McIntyre's fellow architect Gustave Eiffel also built something—Eschol wines again performed beautifully, winning awards and inspiring curiosity, if not yet full-blown respect, for the wine industry in the United States.

N. B. (USE PENCIL FOR MAKING UP STATEMENT.)

Assessment List of the County of Napa, 1890

All property must be assessed at its "full cash value." The terms "value" and "full cash value" mean the amount at which the property would be taken if taken in payment of a just debt due from a solvent debtor. If any person, after demand made by the Assessor, neglects or refuses to give, under oath, the statement herein provided for, or to comply with the other requirements of this title, the Assessor must make an estimate of the value of the property of such person; and the value so fixed by the Assessor must not be reduced by the Board of Supervisors.

STATEMENT OF PROPERTY BELONGING TO, IN THE POSSESSION OR UNDER THE CONTROL OF

Josephine Tychson

At twelve o'clock Meridian, on the first Monday of March, 1890.

PREPARED AND REQUIRED BY THE STATE BOARD OF EQUALIZATION.

(GIVE VALUE OF MORTGAGE IN FIGURES.)					NO.	VALUE
I hold Mortgage against				Brought Forward,		5 3 0
				Cows, Thoroughbred		
				Cows, American	1	2 5
Mortgage against me for $ *1,550* given to				Calves	4	4 1
H. B. Martin, San Francisco				Stock Cattle		
two thousand St. Helena				Beef Cattle		
Bonds, State, County or Municipal *None* $ yr.				Goats, Cashmere		
Unsecured Credits, Notes, Dues, and Demands owing				Goats, Common		
me by others, including deposits in any bank or firm $				Sheep, imported of fine		
Unsecured Debts, due to bona fide residents of this				Sheep, Common		
State deducted from above $				Lambs		
Balance assessable, carried down $				Poultry dozen	3	1 5

	NO.	VALUE						
		Hogs				
Watches		Grain tons				
Jewelry and Plate		Hay tons				
Furniture	5 0	Wood cords				
Piano, $ *1.50* Organ, etc., $	1 0 0	Lumber M				
Sewing Machines	2 0	Wine *1,500* gallons			8 0 0
Firearms		Brandy and other Liquors gallons				
Libraries		Casks, Oak, oval gallons				
Merchandise, of what general kind		Tanks and Puncheons *25,000* gallons			5 0 0
		Hops lbs.				
Consigned Goods		Wool lbs.				
Fixtures of Saloons, Stores, Offices, etc.		Coal tons				
Furniture and Safes in Offices		Boats and Water Craft				
Farming Utensils	2 0	Quicksilver in flasks				
Machinery	4 5	Quicksilver, Flasks, empty				
Wagons, Buggies and other Vehicles	1 5 0	Property held by me in trust				
Harness, Robes, Saddles, Blankets, etc	4 5	Separate Property of wife, including Jewelry				
Horses, Thoroughbred		Property of minor children				
Horses, American	1	1 0 0	Any other personal property, describe it				
Horses, Halfbreed						
Colts		Total value exclusive of money and credits			1 6 1 0
Mules		Money on hand, or Special Deposit where and with whom				
Jacks and Jennets		Solvent credits brought down				
Forward,	5 3 0	Total Value of all Personal Property				

Assessment list of the County of Napa, 1890, signed by Josephine Tychson, which hangs at Freemark Abbey. *Photo by author.*

Also in 1889, yet another German immigrant to the Napa Valley, John Henry Fisher, built yet another winery. Fisher built his on the edge of the crater of an extinct volcano high in the Mayacamas. He named the winery Fisher and Sons and dug a small cave into the stony earth. Fisher transported barrels of wine down the treacherous mountain trails, using a horse and wagon, to the banks of the Napa River, where the wine was then ferried to San Francisco and sold. When the earthquake of 1906 destroyed his other assets, Fisher was forced to sell his winery and the surrounding 250 acres to the Illinois Pacific Glass Company. The property soon found its way into the hands of Pietra and Maria Marenco, who most likely fancied the isolated location as an outpost for bootlegging.

In 1895, the Salmina family from Switzerland, who had arrived in the Napa Valley decades before, leased Larkmead. The solitude of the valley had failed to satisfy Lillie Hitchcock Coit, now nicknamed Firebelle Lil, who spent the rest of her life between Europe and San Francisco. Felix Salmina, who had some background in winemaking, quarried stone from the Vaca Mountains and, as he transitioned from lessee to owner, built himself a winery on the property. Not long after, and not far south, the aging Josephine Tychson sold her winery in 1898 to Antonio Forni of Lombarda, Italy, who renamed the estate Lombarda Cellars. Forni tore down Tychson's redwood

Wooden gravity-fed winery designed by Hamden McIntyre for the Eschol brothers, today the home of Trefethen. *Courtesy of Trefethen Family Vineyards.*

Terraced vineyards at Mayacamas, originally planted by John Henry Fisher. *Courtesy Mayacamas Vineyards.*

structure and replaced it with a large stone winery. All around the valley, more and more structures were going up. As more and more people moved north of San Francisco, it seemed that the chance to dwell and labor in paradise was just too irresistible for many to pass up.

IT HAD LONG BEEN held that European grapevines, *Vitis vinifera*, produced superior wine to those vines native to North America, so when vines in the southern Rhône Valley, a storied French region that spans from Côte Rôtie in the north to Châteauneuf-du-Pape in the south, began shriveling up in 1863, it caught the attention of the world. The source of the trouble was soon traced to tiny bugs called phylloxera, which demonstrated no qualms about departing the Rhône Valley and quickly spread across the vineyards of France. By 1869, phylloxera had infested well over half of Bordeaux's 170,000 hectares of vineyard, and the minute pests quickly turned their unslakable appetite toward the Loire, Champagne and beyond. Hungry, winged, these destructive parasites then spread across Europe, and in the decade that followed, phylloxera ravaged the European wine industry. Estimates suggest that between 70 and 90 percent of grapevines were destroyed, and European wine production was decimated.

Insects of the order Hemiptera, the suborder Sternorrhyncha and the family Phylloxeridae, phylloxera proved truly devastating. Stevenson, who was touring California sometime after the phylloxera outbreak in his native Europe, introduces the third chapter of *The Silverado Squatters*, entitled "Napa Wine," with lamentations about the destruction. "The unconquerable worm invades the sunny terraces of France, and Bordeaux is no more, and the Rhône a mere Arabia Petraea. Château Neuf is dead, and I have never tasted it; Hermitage—a hermitage indeed from all life's sorrows—lies expiring by the river." Stevenson's depiction is as grim as it is poetic, but it is hardly hyperbolic. There was little reason to think that the "unconquerable worm" would fail to triumph; the epitaph of the great French châteaux had already been sent to the printer. "It is not Pan only; Bacchus, too, is dead," concluded Stevenson.

By the time phylloxera was first discovered in California, at Agoston Haraszthy's vineyard at Buena Vista, most of Europe's vineyards lay in shambles. Toro, in northern Spain, was one of a few exceptions. There, a natural amalgamation of sandy soils made up of limestone, clay and puddingstone was enough to keep the destructive pests at bay. Replacing the soils of Bordeaux and Sonoma with sand like that found in Toro, however, was out of the question. While winemakers, vineyard owners, government officials and scientists scrambled to find a solution, the budding American wine industry was forced roughly to its knees.

After much trial and error, the elusive solution was found in grafted rootstock. In short, vinifera could be grafted onto American rootstock, such as Rupestris St. George, creating a vine that was European above ground but American below and, thus, resistant to phylloxera. This process of grafting allowed native European varietals such as Cabernet Franc, Pinot Noir, Sauvignon Blanc and Chardonnay to grow with little threat of being infested by phylloxera. In Europe, replanting was expensive, labor-intensive and time-consuming, as was the process of getting Europeans to accept the solution. Though it became legal in France to perform such grafting in 1878, the Bordelais did not begin until 1881, and it took the stubborn Burgundians until 1887 to do so.

The United States had the benefit of observing the Europeans' struggles before proceeding with its own grafting and replanting operations. Soon, the wine industry in California boomed again. In the Napa Valley, the three thousand acres of vineyard that had not been lost to phylloxera multiplied six times, while the number of wineries soared from low double digits early on to as high as 166. In 1885, an Italian doctor, Osea Perrone, purchased 180 acres in the Santa Clara Valley, in the Santa Cruz Mountains south of San Francisco. Perrone terraced vineyards near the top of Monte Bello Ridge and called his winery Monte Bello. In 1892, Perrone produced his first vintage. All across California, the wine industry was showing promise once again. The disease, it seemed, had been defeated, and the excitement surrounding wine in America could continue.

THE SCENE WAS GRIM and melancholy in Yalta, a city on the southern tip of Crimea that juts southward into the Black Sea, creating the peninsula's distinct silhouette. The bitter winter of 1920 had set in. Viktor Tchelistcheff, a prominent member of the Russian aristocracy—which had crumbled when Tsar Nicholas II abdicated his throne in the midst of the Great War—fled to Crimea with his family to avoid the ire of Russia's new communist rulers. He was a "class enemy," and he and his family had no future in post-Romanov Russia. At great expense, they had managed to escape, and Viktor's son André had enrolled in the military academy in Kiev, where he became a second lieutenant and joined the White Army to battle Leon Trotsky's Red Army in the struggle for Russia's future.

Trotsky had all the advantages a military commander could hope for, not least of which was significant numerical superiority. When André Tchelistcheff's unit attacked a Red Army outpost heavily fortified with machine guns, there were few survivors. André Tchelistcheff was left for dead on the frozen, snow-covered battlefield. Back in Yalta, André's parents and siblings held a funeral for him in the Russian Orthodox church in which young André had served as an altar boy growing up.

Despite the deep grief of his parents and what must have been an elaborate funeral, given his family's station, André Tchelistcheff was reluctant to die. When a group of Cossacks arrived to remove the dead from the battlefield, André was discovered, badly frostbitten and hypothermic but very much alive. He was taken to nurses who, for more than two months, worked to keep him as such, massaging his frostbitten legs in olive oil and feeding him what they could. In time, his strength returned.

André Tchelistcheff was born in Russia in 1901. The year before, half a world away, fate had begun spinning a most elaborate web. Among the new wineries to spring forth during the era of rebirth in the Napa Valley was the love child of Fernande and Georges de Latour. At the beginning of the twentieth century, the couple purchased a modest four acres of land and named it Beaulieu, from the French *quel beau lieu*—"What a beautiful place!" They then purchased the defunct Ewer and Atkinson Winery and reimagined it as the perfect complement to their vineyards. Carefully, they planted their land with vines from France that de Latour was convinced would be resistant to phylloxera. They couldn't have known it in 1900, but the fruits of this labor would ripple across the Napa Valley for generations to come, when their paths would merge with that of André Tchelistcheff.

EUROPEAN IMMIGRANTS CONTINUED TO flow into the United States, attracted by the opportunities they saw. Both land ownership and upward mobility were in short supply in Europe, far more available in the fledgling Americas. Giuseppe Pelissa immigrated from the Italian Riviera to San Francisco in 1902, where he joined relatives, the Carillos. Benicia Carillo had married General Mariano Guadalupe Vallejo and made a comfortable life for herself and her family. Giuseppe had grown Vermentino for personal consumption back in Italy. He married Mary Louise Pietronave in 1903, and soon, the couple moved to Calistoga, where they purchased thirty-five acres of riparian land with a ten-dollar gold coin. Soon, Giuseppe, who by then went by Joseph, had planted a vineyard.

On a similar path, another Giuseppe named Simi and his brother, Pietro, immigrated from Montepulciano and started their eponymous winery in nearby Sonoma. In 1904, however, both died within a few weeks of each other, leaving Giuseppe's eighteen-year-old daughter, Isabelle, to run their winery. Soon after, a nineteen-year-old Italian immigrant from Genoa made a batch of wine in San Francisco. His father was so impressed by what he tasted that he sent the teenage Louis M. Martini back to his home in Italy to formally study winemaking.

The University Farm Bill called for the addition of a farm to the University of California land grant institution, and Davisville, the rural township established half a century before by Jerome and Fanny Davis and Joe Chiles, was selected as the location. In 1907, the town's name was shortened to Davis, and not long after, buildings began to go up. Originally a mere extension of the Berkeley campus, this institution would eventually become one of the most important viticultural laboratories in the world. With phylloxera in the rearview, California's wine industry was free to flourish again.

The Great War began in 1914, though it would take the United States more than two more years to enter it. American wine was selling, due in part to a shortage from war-torn France. True to form, however, Napa was not the only beneficiary of the boom. Even farther south than Osea Perrone's winery in the Santa Cruz Mountains, Monterrey County also played host to a growing viticultural scene. In one two-year period in the middle of the nineteenth century, the number of planted vines in Monterrey multiplied fivefold, from ten thousand vines to fifty thousand in that short span of time. Nearby Soledad was a thriving community and rapidly growing. The Pinnacles, haunting rock formations caused by volcanic activity and seismic shifts millions of years before, were a natural curiosity that attracted tourists to the area. President Theodore Roosevelt had declared them a national monument in 1908. As the area grew, the viticultural industry in the area also expanded.

By 1919, in what now seems like a masterstroke of bad timing, a Frenchman named Charles Tamm had planted Chenin Blanc on a benchland high above the valley floor. Tamm taught French in San Francisco while working for the Belgian Consulate and had little time to tend his vineyard. On the adjacent property, the Dyer family had a ranch where they hunted game and grew grapes, but the Dyers were teetotalers and grew only table grapes. It would be some time before this area grew to meet its full potential as part of wine country.

Having helped to secure peace in Europe, U.S. soldiers returning from "over there" found themselves coming back to a country again at war, this time with itself over the issue of alcohol. The extensive damage that had been done to vineyards and those who tended them by the flaxen, parasitic, aphid-like insects was not to be the final plague, and the wine industry was up against an even more aggressive foe than the phylloxera that had nearly consumed it. The second wave of plague to hit the wine industry was man-made, and as usual, humanity proved to be a far more destructive force than anything Mother Nature could have summoned. When Nebraska became the thirty-sixth state to ratify the Eighteenth Amendment on January 16, 1919, the deed was done. Wineries, distilleries, breweries and bars were ordered to close their doors for good. The University of California Department of Viticulture and Enology obligatorily discontinued operations. The "noble experiment" had begun.

In the thirteen years that followed, the alcohol industry in the United States would go underground, but not away. During that time, American appetites for liquor would create a black market industry that, all told, garnered some

The Pinnacles volcanic formations near Charles Tamm's vineyard, which would become Chalone Vineyard. *Photo by author.*

$36 billion and greatly strengthened organized crime syndicates. American prohibitionists and the politicians who acquiesced to their demands had paved the road to hell with their good intentions.

During Prohibition, American consumption of wine rose from 60 million gallons per annum to 150 million, but it also transitioned from fine, well-crafted wines made by professional winemakers to those that could be concocted by unskilled would-be vintners in makeshift fermentation vessels using grapes that had been shipped unceremoniously across the country in train cars as fruit or grape juice concentrate. Where previously the skilled European immigrants in the Napa Valley had crafted elegant wines on the estates where the grapes had been grown, now Sam in Skokie and Paul in Poughkeepsie had to ship grapes thousands of miles in unrefrigerated train cars, crushing them with shovels if they felt pragmatic, their feet if they felt romantic, before sloshing the juice, along with yeast if they could find it and copious amounts of sugar, into a lead clawfoot tub they had set up in the garage, covering it with a musty blanket to keep the flies and mice at bay. For this, Sam and Paul preferred Alicante Bouschet, not Cabernet Sauvignon, as the former have thicker skins that could impart more tannin, flavor and color and were somewhat better suited to enduring the harrowing train ride. Rapidly, the American palate devolved.

This phenomenon of outlawing professional viticulture while allowing home winemaking to remain legal had a profound impact on the American wine industry in two ways. In Napa and the rest of California wine country, it often meant tearing out vineyards of fine *Vitis vinifera* and replacing them with grapes of inferior quality. Those vineyard owners who didn't do this often tore out their vines anyway, planting a variety of fruit and nuts in the fertile soils instead. Halfway through Prohibition, there were more plums than there were grapes growing in the Napa Valley.

The second manner in which the circumstances of Prohibition impacted the wine industry was even more damaging in the long run. As previously stated, consumption more than doubled during this time, but that consumption was of a vastly inferior product, and people soon got used to it. Home winemakers didn't necessarily get better at making wine over time, but they did learn to choke down their own homemade swill, and over the years, the American wine consumer gradually forgot what good wine smelled and tasted like. By many accounts, it would take generations to reverse the damage done to the American palate by the voracious consumption of cheap, sugary, homemade plonk during the ignoble experiment.

At Chateau Montelena, Alfred Tubbs continued to raise and sell grapes with the assistance of his grandson, Chapin, but of course, he was not permitted to make wine. Down the road at Lombarda Cellars, Antonio Forni saw no alternative but to shutter his winery. John Fornay at Chateau Chevalier, so near Miravalle, sold his grapes to the Christian Brothers, a Roman Catholic order that made wine in Hamden McIntyre's impressive Greystone and to whom the government had granted a dispensation for making sacramental wine. While some in the Napa Valley followed Tubbs's example and attempted to make a go of it as grape farmers, a far greater number took Forni's route, closing down their wineries and either letting their grapevines be overrun by natural flora or else tearing them out and replacing them with fruit and nuts for which there was still a market.

The Salminas were in the former category and managed to retain Larkmead by selling fruit to home winemakers and producing sacramental wines. James Fawver, the son of a pioneer, had assumed management of the Eschol Winery and began the laborious task of replanting the vineyards that had been devastated by phylloxera. When Prohibition hit, Fawver continued making wine, convinced that repeal would soon come. When it didn't, Fawver stopped making wine and devoted his energy instead to farming grapes.

Back at Spring Mountain, Parrott had died of stomach cancer in 1894 at the age of fifty-four. Miravalle would sit uninhabited; the hardwood floors were repurposed by local children as a roller-skating rink. Other imposing and important Napa Valley structures, the vestiges of great wealth and monuments to viticulture, were soon to suffer a similar fate. Captain Niebaum had died in 1908, and during Prohibition, his Inglenook Winery sat empty and unused. As Prohibition took hold and the nation's drinking went underground, such ostentatious aboveground winemaking operations were mostly forced into obsolescence, boarded up and abandoned.

While the immigrants who had settled the land to erect their monuments to viticulture—Germans, Spaniards, Finns, Hungarians, Mexicans, Italians and more—all struggled in their own right to survive the beginning of the twentieth century, the Indigenous populations of the Napa Valley fared even worse. Upon his arrival, George Yount had estimated that eight thousand Natives lived in the Napa Valley. By the early twentieth century, fewer than one hundred remained. The 1910 census reported only seventy-three Wappo in Napa. In so many ways, the "settlement" of this exquisite and bewitching valley is yet another story of tragedy for those to whom it rightfully belongs.

In what appeared to be a counterintuitive gamble, if not a fool's errand, three years after Prohibition started, Louis M. Martini returned to California from Italy at the age of thirty-five, by then a skilled winemaker. Though Italy was not dry, Martini made the decision to make wine in the Napa Valley during Prohibition, focusing on sacramental wine and joining many of his contemporaries in producing grape concentrate for home winemakers.

Up in the Mayacamas, Fisher's winery was acquired from the Marencos by the Brandlin family in 1921. Though this was seemingly a bad time to acquire a winery, it is rumored that throughout Prohibition this isolated mountain fort remained a hub of bootlegging activity. While some continued to fight valiantly to make ends meet, whether legally or otherwise, many at that time left the industry for dead.

JUST AS ERIS'S GOLDEN apple sowed the discord needed to spark a war that would, all told, bring down an entire proud civilization to such an extent that, in the distant future, there would be serious scholarly debate about whether or not the city of Troy had ever even existed, the relentless persistence and faux piety of the prohibitionists instilled fear and stoked the hatred necessary to bring about the end of an era, to decimate the livelihoods of countless civilians and to destroy an ancient and noble form of art. There was no doubt that the Napa Valley possessed the beauty of Aphrodite, but it could not endure the wrath of slighted Hera and Athena. And so, just as quickly as the area had demonstrated its tremendous, possibly unparalleled, potential to become one of the great wine-growing regions of the world, that world tore these virtuous vines up by their roots and tossed them on the burn pile.

Chapter 2

PREPARING FOR WAR

Without a sign, his sword the brave man draws,
and asks no omen, but his country's cause.
—The Iliad

G eorges de Latour was feeling, if not outright confident, then at least cautiously optimistic. He and his son-in-law had taken a detour from the family's vacation in his native France to the Institute of National Agronomy in Paris, where de Latour was hopeful that Paul Marsais, the renowned enologist, might agree to serve as winemaker and problem-solver-in-chief at his Beaulieu winery in Rutherford, California. Marsais might have found the idea of leaving Paris for rural America amusing, but he was far from persuaded. Marsais suggested his star pupil, a brilliant researcher who spoke six languages, had an uncanny knack for viticulture, made up for his diminutive stature with the energy of two ordinary men and was idealistic almost to the point of naivete. The young man had taken the scenic route between Russia and France but wasn't overly forthcoming with additional details about his past. When Paul Marsais introduced Georges de Latour to André Tchelistcheff, de Latour offered him the job on the spot. Tchelistcheff, urged on by Marsais, accepted.

André Tchelistcheff had lived several lifetimes since his funeral was held in Yalta some eighteen years prior. He had spent over a year with his fellow soldiers in an internment camp in Gallipoli, done hard labor in Bulgaria and fought as a mercenary. Through a program for Russian refugees with some

Andre Tchelistcheff examines a grapevine at Beaulieu. *Courtesy Napa County Historical Society.*

level of education, he had been admitted to the Institute of Agricultural Technology in Brno, Czechoslovakia, where the aspiring MD reluctantly accepted a role studying animal physiology instead. Part of his studies included six months in Hungary, in Tokaj, the young Russian's first real exposure to viticulture and the production of the wines that his aristocratic father had so enjoyed before the revolution. Soon after, he took a job as an agronomist in Dubrovnik, Yugoslavia, which furthered his exposure to wine. When Tchelistcheff informed his wife, another White Russian refugee named Catherine, that he had accepted the job and they would be moving to the United States—to California, of all places—she began to weep.

After a journey that took them over the ocean and across the seemingly boundless expanses of their new home country, André, Catherine and Dimitri Tchelistcheff arrived in San Francisco in September 1938. When he and Catherine at last arrived in the Napa Valley, the man who had once been left for dead on a Crimean battlefield looked from the Mayacamas to the Vacas, from Mount Saint Helena to San Pablo Bay, and concluded that he was standing in "the most beautiful place I'd ever seen."

JACK AND MARY CATHERINE Taylor had first visited the Napa Valley in 1936, staying on Mount Veeder at the Lokoya Lodge. They fell in love with their surroundings, and Jack, an Englishman and an executive at Shell Oil, vowed to return. The couple did so in 1941, purchasing a defunct old winery and distillery on Mount Veeder from Henry Brandlin. The winery had belonged to a man named John Fisher, among others, and boasted some of the oldest wine caves in the Napa Valley.

The Taylors attempted to plant grapes shortly thereafter, but they were unable to keep deer out of their vineyards. A war was going on, and the necessary supplies were rationed. The Taylors renamed their winery Mayacamas, after the mountains in which it sat, and eventually succeeded in fencing their vineyards and planting Cabernet Sauvignon and Chardonnay. For some time, the Taylors managed their investment from a distance, but in 1945, they and their three children moved onto the property, into the old distillery, which they converted into a home.

ROBERT MONDAVI LOOKED PEACEFULLY out the passenger side window of his father's car, relaxed and at ease as the noontime sun warmed the vinyl beneath him and tempted him to doze off. It would have been impossible to

Jack Taylor out in the vineyards at Mayacamas. *Courtesy Mayacamas Vineyards.*

sleep, however, despite the rhythmic lullaby the engine was softly humming, given how inspired he felt. His brother, Peter, sat in the back seat. Peter was the youngest of four, and when Robert was in the car, Peter's seat was in the back. Like Robert, Peter had attended Stanford and, like Robert, graduated with a degree in economics. After Stanford, Peter began studying enology at Berkeley, but his studies were interrupted by the Second World War. Home on a brief furlough, Peter agreed to accompany his father and brother on a drive north from their home in Lodi.

In the spring of 1943, America was embroiled in a two-front war. Twelve thousand acres of the Napa Valley remained planted to grapevines, roughly the same amount as had been under vine at the dawn of repeal in 1933, and while prunes were still viewed as the superior cash crop of this fertile breadbasket, the Mondavis recognized the potential for a return to viticulture. Two nights before, Robert had an audience with his father, Cesare. Louis

Left to right:
Robert, Cesare
and Peter
Mondavi.
*Courtesy Joyce
Stavert.*

Stralla, who had been leasing a historic winery in the Napa Valley, tipped Robert off that it was up for sale. Robert was determined to purchase it, but Cesare was unmoved by Robert's initial pitch. "I'm happy," Cesare told his son gently. "I don't have to get any bigger than what we are." Cesare was content with his business of shipping wine grapes from Lodi and keeping an eye on his Napa Valley investment, the Sunny St. Helena Winery, which he had bought into six years earlier. Rosa, Robert's mother, however, overheard the conversation, and after Cesare went to bed, she interrogated Robert, who, ever the salesman, convinced her it was a good investment. The next morning, Cesare emerged from their bedroom and asked Robert when he'd like to view the property. When Peter arrived home, the three men were off, driving north into the Napa Valley.

From the front seat, Cesare took in the lush orchards of fruit and nuts. He deftly maneuvered their vehicle along the familiar, rolling roads, through the small city of Napa and then past the tiny, nearly uninhabited Yountville. They traveled through St. Helena, for which the family's existing wine investment was named. Just outside of town and to the west, the Rhine House, the Beringer brothers' old estate, stood amid trees that were, by now, as tall as it was. During Prohibition, the Beringers made sacramental wine, and after home winemaking ruined the American palate, the Beringer estate capitalized by making sweeter wines and offering tours on the weekends.

Robert and his wife, Marjorie Ellen, had taken the tour at Beringer once and enjoyed themselves. Like his Italian immigrant parents and brother, Robert didn't particularly enjoy the sweet jug wines for which the Napa Valley was then known. The Mondavis realized that between these two mountain ranges lay the potential to produce truly excellent wine, and they were setting out on a quest to unlock it.

Not far past Beringer, the imposing walls of Greystone towered in the shadows of the Mayacamas. Across the road, Robert directed Cesare to turn down the straight, tree-lined drive, toward an ancient stone building. "It has good bones," the realtor had said to Robert on a previous visit, the sort of thing realtors habitually utter when they are attempting to excuse the outward appearance of a place. The good bones supported what was left of Charles Krug's winery. Cesare enjoyed the enthusiasm on the faces of his sons, and more importantly, they seemed to agree on this idea. Cesare thought their consonance a nice change of pace from the ordinary brotherly bickering. He parked the car. They bought the winery. The rebirth of the Napa Valley had begun.

Not long prior to the Tchelistcheffs' arrival, the University of California reestablished its Department of Viticulture and Enology on the Davis campus. Prohibition had forced a break in the action, but that was in the past, and now it was time to get back to the business of enology. The university's investment in the industry was a triumphant signal and an even more important resource for aspiring viticulturalists. In the decades that followed, UC Davis would train many of the winemakers who would bring wine back to life in the United States. One such was MaryAnn Graf, who graduated with a degree in fermentation sciences in 1965. The first woman to earn such a degree, Graf picked up where Josephine Tychson had left off, pioneering a role for women in wine country. Graf's first move on graduation was to take a job at Simi, making her the first woman to become head winemaker at a major California winery in the post-Prohibition era.

Even his broad spectacles couldn't help Jack Davies find a reason to remain in Los Angeles. The city was changing, not for the better, and he was entering middle age. Jack Davies was a successful businessman, but the noise and sprawl of Los Angeles had grown tiresome, and at the age of forty-two, he was ready for a change of scenery. Seeking a more rural setting, Jack and his wife, Jamie, a slender, elegant woman with a winning smile to match her husband's, toured one dilapidated Napa Valley ruin after another, until at last, they found a run-down Victorian mansion in the shadow of Mount Saint Helena with overgrown gardens and dank wine caves dug so deep into the mountainside that a person could get lost in them. The house was full of dust, the caves full of bats and the couple full of irrational optimism and intrigue at the idea that this property, once belonging to a German barber who had famously played host to a Scottish author, would now become their own. The couple threw their hearts and souls into restoring Jacob Schram's ancient house and winery. The Mondavis hadn't listened when they were told they couldn't make exceptional wine in the Napa Valley, and they proved doubters wrong. The Davieses were similarly disinclined to listen when people scoffed at their plans to make sparkling wine and set to work with the intent of rivaling the famed cuvées of Champagne.

Though the practice was far from commonplace in the 1940s, Beringer was not the only winery in the Napa Valley open to visitation. Albert "Abbey" Ahern, Charles Foster and Mark Freeman were friends who found the stone exterior and quaint Italian character of Lombarda Cellars charming. They purchased Lombarda Cellars, formerly Tychson Cellars, and bestowed on it a third moniker, dissecting their names and reassembling them as Freemark Abbey. Sorry, Charlie.

Joe and Alice Heitz were another couple who took note of what Cesare, Rosa and their family were doing on the old Charles Krug estate and became early and important participants in the Napa Valley's renaissance. A farm boy from Illinois, Joe was studying veterinary medicine when World War II broke out, calling him to the U.S. Air Force, where he was stationed in Fresno. After the war, Joe decided to stay in California and study enology at UC Davis. On his graduation, Heitz worked a brief time for Ernest and Julio Gallo but soon succumbed to the pull of Napa, where he worked alongside André Tchelistcheff at Beaulieu Vineyards for seven years. Heitz was grateful for the opportunity to learn alongside a master, and while the pair disagreed about whether it was best to fine using egg whites or gelatin, they liked one another.

Eventually, Heitz left Beaulieu to help launch the enology program at Fresno State. Not long after, Joe and Alice purchased a small parcel of land south of St. Helena planted to Grignolino. They started a winery and, in the early years, purchased wine from Christian Brothers Winery, then the lone occupant of the titanic Greystone building, putting their own label on it. Soon, Joe put his education to work making his own wines. Joe and Alice asked their ten-year-old son, David, to design their label, and Joe set out to make wines of the same quality as those he and Tchelistcheff had made at Beaulieu.

The post-Prohibition rebirth of the wine industry continued. Not long after de Latour met Tchelistcheff in France, a theologian by the name of William Short purchased an abandoned mountaintop winery south of San

Francisco. He tore out the overgrowth on the terraces and replanted the old vineyards to Cabernet Sauvignon. Short would eventually sell the winery to a group of engineers from Stanford. Like so many other historic wineries around this time, Osea Perrone's long-lost vision had been given new life.

Farther south, in Monterrey, Will Silvear had assumed control of Charles Tamm's operation and was busily planting Chardonnay and Pinot Noir in the limestone-based soil that reminded him, like his predecessor, of Bourgogne. Silvear sold grapes to Tchelistcheff at Beaulieu until his death in 1955. After his death, Silvear's wife briefly carried on the business. When she ultimately sold the property, it would change hands repeatedly. In 1960, it was given the name Chalone in tribute to a local Indigenous tribe by then-owner Philip Togni, and while the property continued to change hands, the name stuck. Not long after, Dick Graff came to own the property. A Harvard graduate who attended UC Davis to learn to make wine, Graff repurposed a chicken coop on the property into a winery and soon began selling wine.

Back north of the bay, the Napa Valley continued bustling with activity. Chapin Tubbs, grandson of A.L. Tubbs, rebranded his grandfather's winery as Chateau Montelena, an homage to the mountain in whose shadow the winery dwelt. Chapin Tubbs passed away, and in 1958, the family sold the

The chicken coop originally repurposed as the winery at Chalone Vineyards. *Photo by author.*

winery to Yort and Jeanie Frank. The Franks had emigrated from Hong Kong prior to the war. They dug a pond, filled it with koi and installed Chinese-style gardens on the grounds. Like Chapin Tubbs's, however, their time as the caretakers of Chateau Montelena was to be brief. The number of people who wanted to own and operate a winery, it seemed, far exceeded the number who had the ability, resources and gumption to execute the idea.

ANDRÉ TCHELISTCHEFF TOOK A slow drag from his cigarette and looked thoughtfully into the intelligent eyes of the equally diminutive man seated in front of him. Wine was not the only thing these two globetrotters had in common. Both had backgrounds in agriculture, and both were highly educated and equally gifted. Much like Tchelistcheff, the man across the table from him had come to California "the long way" en route from his home in Croatia. Tchelistcheff had spent time in Zagreb, and both men had fled Europe as refugees in search of a better life in the United States. Miljenko Grgić, who early on changed his name to Mike Grgich, was self-conscious about his English and studied hard to improve it. Tchelistcheff sensed this and, switching languages without warning, began to speak to Grgich in his native Croatian. Grgich smiled widely across the table at this strikingly kindred spirit. Tchelistcheff, it seemed, was a godsend.

And yet the twosome had no shortage of differences, either. Tchelistcheff was born into the aristocracy before circumstances deprived him of the comfortable future that awaited him in czarist Russia. Grgich, on the other hand, was born into a large family with very little money and no modern amenities in their home. The youngest of eleven children, Grgich was born in Desne, a hamlet of around one thousand people on the west bank of the Neretva River, which flows into the Adriatic. His mother made his shoes by hand from animal hide. While the Russian's earliest exposure to wine was observing his father's luxurious partaking in their grand estate, the Croatian's came at the age of two and a half, when his mother switched him from breastmilk to *bevanda*, a mixture of half wine, half water, meant to make the water from their cistern safer to drink. "Someday, you might thank me," the doting Croatian mother told her distraught little boy at the time. "I have liked wine ever since," remarked Grgich nearly a century later.

Grgich had heard California described by one of his professors as "paradise" and hoped one day to move there. Having forgotten his umbrella one day while still in Croatia, he purchased a blue beret to keep his head dry. Now, in sunny California, Grgich removed his beret as he shared his story with the Russian across the table. Only a few months before graduation, he told Tchelistcheff, one of his professors came under fire for openly believing in God, a crime of which Grgich was also guilty. This offense, coupled with the professor's criticisms of communism, had him ousted from the university two months prior to retirement. An incensed Grgich and other classmates complained of this injustice, and Grgich soon found himself being followed by the secret police. So close to earning his diploma, the talented young Croatian had thirty-two dollars—all the money he had saved in his entire lifetime—sewn into the sole of his shoe, filled a cardboard suitcase with enology textbooks and fled for his life.

Like countless others, Grgich and Tchelistcheff were victims of violent communist takeovers. Each man had his own odyssey, and each came to America through other nations, the Russian via France and the Croatian via Germany and then Canada. Both men had chosen California and the Napa Valley.

Grgich's first stop in the valley was working for Lee Stewart at Souverain. Stewart did Grgich a service by offering him the job that got him into the country, but there was little opportunity to advance in the position. One day at Souverain, Grgich was surprised to notice the symmetrical, maplelike leaves of Crljenak Kaštelanski growing in the vineyard. He learned that the locals called it Zinfandel, but he knew the grape came from his native Croatia. It was rumored that young Aldo Biale was making the stuff on the valley floor without a license, selling it in jugs to help support his family. Stewart didn't think much of this, but Grgich didn't mind; he knew what it meant to fight to survive.

Grgich left Souverain Cellars and was hired to work in sparkling wine production by Brother Timothy at Christian Brothers, but again, there was little opportunity to advance. Staring across the table at Grgich, Tchelistcheff felt in some ways as if he was looking into a mirror. He recognized Grgich's abilities just as readily as he could relate to his struggles and hired him to serve as his assistant enologist at Beaulieu in November 1959.

Chicago, Illinois, was a hell of a place for an immigrant family to try to carve out a life—but then, wasn't all of America? While Al Capone was busy making a fortune on Prohibition and establishing organized crime syndicates that would last into the twenty-first century and beyond, Stephen and Lottie Winiarski were welcoming a baby into the world. In their native Polish, the surname Winiarski means "of a winemaker," and Stephen busied himself regularly by fermenting everything he could find, from honey to fruit to dandelions. The baby was named Warren, and as a boy, Warren Winiarski held a curious ear up against his father's barrels, listening to the odd yet inviting churns and gurgles of fermentation. Warren's curiosities didn't end there, however, and he would soon leave Chicago to study at St. John's, where he met Barbara, who would become his lifelong companion.

Warren and Barbara returned to Chicago, briefly, where the former completed a graduate degree in political science, before being drawn to Italy to research political theory. Reflecting later in life on his journey into the world of wine, Winiarski commented that he never thought wine would become, for him, what it did: "I couldn't imagine when I was in Italy doing work on Machievelli's writings…except Italy became wine." As a scholar abroad, Winiarski took advantage of the local culture. "Since I didn't have lots of funds, I sat in with students and we shared wine together, usually out of an unlabeled bottle somehow," he later recalled. "Wine was everywhere. It was so pleasant." In Italy, Winiarski was exposed to wine in a way he had never been before. Not only was wine partnered with meals, but he also discovered that Sangiovese, Nebbiolo and Primitivo grapes were far superior to dandelions and honey in terms of what the finished product could become. "I carried all of that back to Chicago when I left Italy," Winiarski said.

Warren and Barbara first returned to Chicago, but wine had managed to grow tendrils around the couple. One night, a friend came to dinner and brought with him a bottle of wine from the East Coast. "We shared that wine together and suddenly, that wine *talked* to me," recalled the winemaker later in life, without the slightest hint of hyperbole. The wine said: "*Listen* to me! I'm *talking* to you. I want you to hear what I have to say. I want you to take notice of me." The Winiarskis felt the tendrils begin to tug them westward, toward California and wine country. "Nothing was the same after that," conceded Winiarski. In 1964, they loaded a trailer behind their Chevy and struck out for California. After struggling to cross the country in a car not built to pull a trailer up a mountain, they at last arrived. There was just one problem. "With two children, I couldn't afford to go to school to study winemaking," said Winiarski. No matter, he thought. "I had to learn by doing."

To do so, Winiarski first found work with Martin Ray at his winery in the Santa Cruz Mountains. The two men could have been better suited to work together, however. "He was a sort of despot," recalled Winiarski, who despite this admired Ray. "It was his way or no way. And I wasn't quite that way. Everything was magical when I went there, but there could be this difficulty because he was so single-minded." Winiarski's tenure with

Ray was short-lived. Soon, he and Barbara loaded the children into their much-abused Chevrolet and headed north to Napa, where Winiarski found a second winery where he could learn by doing.

"I found a job as the number two man at a two-man operation," he joked. By some accounts, Lee Stewart made Martin Ray look amiable, but Winiarski admired his wines and wanted to learn everything he could from the man. "I had a steady income and could support myself and my family while I was learning the trade of wine." Over the next two years, the meticulous and methodical Stewart taught Winiarski a great deal about how to make wine, and what he learned would serve Winiarski well throughout his career.

Maria Brambila Dios was, like so many Mexican women in her position, unshakable. Along with her husband, José, and their children, Maria moved from San Clemente, Jalisco, Mexico, to the United States in search of opportunity. While José worked the land, tending to the vineyards at Beaulieu, Maria took care of the children and the home they rented within walking distance of the winery.

Beaulieu is where a young Gustavo Brambila, their son, was introduced to Tchelistcheff, Grgich and other fellow immigrants who made him feel welcome and whose passion and knowledge of wine were such that he could not help but to find it contagious. One day, during crush, Gustavo's father walked across the street from Beaulieu to where he and his family were living, and he brought with him a bottle of muscat juice from work. He encouraged his young son to try a taste of the relatively sweet liquid.

When his father returned to work that afternoon, little Gustavo dutifully attempted to recap the mostly full bottle before placing it in the refrigerator. Two weeks later, the long-forgotten bottle exploded, dousing the precious food inside their Frigidaire with sticky grape juice. "I told my mother at that time that I didn't know what happened, why it happened, but I would find out and I promised her that I would find the reasons and so forth and I would get back to her," recalled Brambila.

Meanwhile, the immigrants at Beaulieu worked hard to impress on José Brambila the importance of education. It was too late for José to go to school, but it was important that he send his children, if he could manage it. Gustavo's search for answers about exploding grape juice, coupled with the good counsel his father was given, would eventually lead him to attend UC Davis and study the science of wine.

WHILE SO MUCH OF what Robert Mondavi accomplished in his lifetime was the result of being a truly forward-thinking visionary, the Robert Mondavi Winery was established out of necessity born of desperation and as the result of an irreconcilable rift in the Mondavi family between Robert and his brother, Peter. When Cesare Mondavi agreed to purchase the historic Charles Krug Winery, it was on the condition that the two brothers run it together. Cesare had been the glue that held the Mondavi family together, the mediating factor between two forceful, intelligent, stubborn young men who often failed to see eye to eye. After Cesare's death in 1959, the brothers managed, as Robert would later put it, to "hold an even keel for a while," but the two seemed fated to do battle, and when it came to each other, it took little to set either of them off.

What ultimately did them in was the now-infamous purchase of an expensive fur coat—or rather, the tensions that simmered afterward. Robert and his wife, Marge, were invited by President Kennedy to represent the winery at a state dinner at the White House. Robert and Marge splurged on a mink coat, in which Robert thought his wife looked "smashing." Prior to the dinner, President Kennedy was assassinated, and the dinner was held later at the behest of President Lyndon Johnson, a "rather subdued but elegant affair" as Robert later recalled. Two years passed, but it seems that Robert's taste for expensive things festered in his brother. One day, at a family gathering, Peter angrily raised the issue of the expensive coat to Robert, who punched Peter twice. The two fifty-year-old men wound up in a fistfight, "acting like kids in a school yard—with terrible consequences," recounted Robert. "The fight split our family in two, and there was no repairing the damage," Robert wrote in his memoir. Soon, the family forced him out of Charles Krug.

Starting a winery from scratch was no small endeavor, even for the well-connected Robert. One afternoon in 1967, he showed up at the home of Andrew Pelissa, Giuseppe's son, now the patriarch of the family. The Pelissas owned a ranch near Yountville and had grown Giuseppe and Mary Louise's original thirty-five acres exponentially, with over five hundred acres now planted to grapes, which they sold to the Mondavis at Charles Krug Winery. Ren Harris, Andrew's son-in-law, was then twenty-six and over for a visit. On his arrival, Robert politely made small talk in his usual charming manner before he got around to explaining that he intended to open a winery and was hoping Pelissa would invest.

"Bob," said Pelissa in a fatherly manner. "Go home and patch things up with your family." Mondavi assured him that he would try but asked that

Pelissa keep his offer in mind and get back to him. After he departed, Ren Harris asked his father-in-law if he was going to invest in Mondavi's new venture. "Bob's a hell of a nice guy," replied Pelissa, "but he's crazy."

Eventually, and with difficulty, Robert Mondavi did scrape together the capital he needed to open the first new winery in the Napa Valley since before Prohibition. He purchased land that included parts of To Kalon Vineyard in Oakville, seven miles down valley from Charles Krug, and opened the Robert Mondavi Winery. Robert took revenge on the family that had voted him out by altering the pronunciation of his well-known surname from "Mon-*day*-vee" to "Mon-*dah*-vee" as a final parting shot.

Robert Mondavi's departure from Charles Krug had, if anything, sparked in the two brothers a mutual desire to succeed autonomously from one another. While Robert worked tirelessly at his eponymous new estate, Peter continued making the best wine he possibly could at Charles Krug under the stewardship of his mother, Rosa. Peter and Rosa took chances on people they believed in. Among them was a man named Nils Venge, a Vietnam veteran and UC Davis grad, whom they hired as vineyard supervisor.

Robert Mondavi was similarly inclined to give people a chance, and he hired his first winemaker away from Lee Stewart at Souverain. Warren Winiarski worked tirelessly during the day and read voraciously about winemaking at night. When he wasn't doing either of these things, he was scouting around the Napa Valley, tasting grapes and wine, looking for the perfect terroir in Napa's diverse range of soils and diurnal ranges in which to grow his own Cabernet Sauvignon.

Winiarski left Mondavi in 1968 to consult for Ivancie Cellars in Colorado, and Mondavi next hired Grgich away from Beaulieu. Grgich admired Mondavi's energy, and Mondavi loved Grgich for his talent. The first wine Grgich made for Mondavi, a 1969 Cabernet Sauvignon, won a blind tasting organized by the *LA Times* by going head-to-head with the likes of Louis Martini, Heitz Cellars and Beaulieu. *Wine Spectator* gave the '69 Mondavi Cabernet ninety-nine points. Grgich basked in the recognition and redoubled his efforts.

In 1970, Grigch was introduced to Zelma Long, who was following in Graff's footsteps and studying enology at Davis. "I would like Zelma to work for Robert Mondavi Winery," Grgich told Long's mother on the telephone. "Over my dead body," came the reply. Zelma at first agreed with her mother but relented after the persistent Croatian's third phone call. "Listen, this is going to be a wonderful opportunity for you to learn about winemaking," Grgich told Long. "So I did," she later said. Zelma Long would spend the next decade at the Robert Mondavi Winery, thankful for the time she spent with Grgich as her teacher. "What André Tchelistcheff was to Bob Mondavi and Louis Martini, Mike was to me," she later shared.

Grgich was happy working with Mondavi, but he also understood that Robert had sons who put a lid on how high he could rise at Robert's winery. When Jim Barrett, Lee Paschich and Ernie Hahn offered Grgich a role at their newly resurrected ghost winery, not only as the winemaker but also as a partner with a small share of ownership, Grgich consulted with Mondavi. "Go," Mondavi told his winemaker. "If it doesn't work out, you can always come back." With Grgich gone, Mondavi replaced him by promoting Zelma Long.

Zelma Long holds a glass of wine. *Photo courtesy Zelma Long.*

Lee Paschich had purchased Chateau Montelena, along with his wife, Helen. Grgich met Paschich in 1971 when the latter attempted to kill the former with a rattlesnake. What actually happened was that Robert Mondavi purchased grapes from Paschich, whose vineyards were hospitable to not only grapes but also reptiles, and Grgich later mused that the rattlesnake must have arrived with the Chardonnay. He killed it with a hammer, buried it under a tree and went back to work.

Jim Barrett left the navy and went to law school in 1948, but he was recalled to active duty when the Korean War broke out. After his second tour of duty, Barrett opened a law office. Handsome, charismatic and a visionary with a strong work ethic, Barrett built his one-man operation into a large law firm, and there he met a real estate developer named Ernie Hahn. Barrett's firm served Hahn, and the two became friends, but Barrett was malcontent and began considering a career change. His criteria: a job that would allow him to "watch the sun come up in the morning and go down at night." Agriculture seemed to be calling to him.

Paschich put Chateau Montelena on the market for an even million dollars. Barrett, a pilot, flew up to the Napa Valley to investigate the property, a pilgrimage that became habitual as his friendship with Paschich strengthened. When he was finally ready to pull the trigger on this massive gamble, Barrett called up Ernie Hahn. He told him all about the winery and his plan to produce wines on par with those they loved to drink. He gushed about the winery, how beautiful it was, and concluded with a pointed invocation: "I can't do it alone." "Let's do it," responded Hahn. They bought Chateau Montelena with the stipulation that Paschich stay on as a third partner. He agreed.

From its inception, the new Montelena was intended to be a Cabernet house. Grgich began work in early May 1972. The first task assigned him by Helen Paschich, Lee's wife, was to design the production facility inside the stone exterior of the winery. Helen was followed shortly by Lee, whose request was that Grgich create a five-year budget for the winery. When Grgich presented the budget to the owners, they nearly choked. "How can it be that we won't make any profit for five years?" they asked, exasperated. "Great wine cannot be made in a hurry," responded Grgich. The determination to produce Cabernet Sauvignon was antithetical to turning a quick buck; red wine, good red wine at any rate, needs time to mature. Half a decade of red ledgers, however, was a nonstarter for such successful and ambitious businessmen, so they asked the obvious question: What can we do to make money *now*? Grgich proposed making white wine, which would take substantially less time to age. With a collective sigh of relief, they gave him the green light.

Grgich began with Riesling and Chardonnay. Jim Barrett quietly harbored reservations. "I didn't quit my day job because I figured this thing could be a huge flop," he later remembered. Working alongside Grgich was Jim's son, Bo, and a few other cellar hands. Grgich put his experience to work, determined to craft the best wine he could, regardless of the varietal. He, Barrett, Paschich and Hahn had high hopes for Chateau Montelena. Nobody, however, could possibly have imagined that their efforts would soon flip the world of wine completely on its head.

PART OF WHAT MADE the Napa Valley so successful and so innovative was the almost complete willingness of the winemakers in Napa to share information with one another. Tchelistcheff arranged a monthly meeting in St. Helena, which he dubbed the Napa Valley Wine Technical Group. Over the years, the group was frequented by Peter Mondavi, Mike Grgich, Warren Winiarski and countless others. There, winemakers met, discussed the vintage, talked about what they were learning through experimentation and shared ideas. The philosophy was, in essence, that a rising tide lifts all boats. While the French approach to winemaking at the time amounted to covering your answers with your free hand so nobody could copy your results, the Americans openly passed notes and shared answers.

"We were all young and enthusiastic and hungry, and highly interactive and cooperative, which led to the acceleration of knowledge and the impact of that knowledge on the wines and the vineyards," remembered Zelma Long.

In 1969, Warren Winiarski met a Napa resident by the name of Nathan Fay and sampled some of Fay's Cabernet Sauvignon. Winiarski recognized it as the type of wine he wanted to make, and soon after, he and Barbara went in with investors to purchase a fruit orchard near Fay's, which they converted to a vineyard. Winiarski's curiosity was matched only by his work ethic, and together, he and Barbara were quick to realize their dream. In 1973, Stag's Leap Wine Cellars officially opened its doors, selling wines made from grapes that were harvested from vines a mere three years of age.

Elsewhere in the foothills of the Vaca Range, and around that same time, a young Bordelais winemaker by the name of Bernard Portet set his sights on the area, as well. Young, talented, attractive and clever, Portet had been tasked by John and Henrietta Goelet with locating a vineyard site anywhere in the world where he could craft the best wines. Portet took the task seriously, visiting Australia, New Zealand, South Africa and numerous countries in South America prior to arriving in the United States. Portet traveled to the Napa Valley and agreed with Winiarski's assessment of the land just under the outcroppings in the Vaca Range known as Stag's Leap.

A single mile separated the land the Winiarskis were planting from that selected by Portet, connected by the Silverado Trail. Eager to begin, Portet purchased Cabernet Sauvignon from nearby SDS Vineyard and Merlot—"I don't remember where I got the Merlot," he chuckled years later—and made the first wine for Clos du Val. A left bank–style blend of 80 percent Cabernet Sauvignon and 20 percent Merlot, the very first wine that Bernard Portet created at Clos du Val was destined to make history.

Despite a lack of broad notoriety, despite having nothing akin to the publicity and support enjoyed by the great châteaux of France and despite limited resources, a general lack of experience and the residual effects of the hangover America was still suffering from Prohibition, the wine industry in California chugged steadily forward. Driven by the pioneering likes of André Tchelistcheff, Lee Stewart, the feuding Mondavi brothers, Louis Martini, Joe Heitz and plenty of others, the Napa Valley and surrounding

Above: Warren Winiarski and Nathan Fay share a glass of Cabernet Sauvignon. *Courtesy Napa County Historical Society.*

Left: Bernard Portet samples wine from a barrel at Clos du Val, circa 1975. *Courtesy Clos du Val.*

areas were alive with the gurgling noises of fermentation that had fascinated Warren Winiarski so greatly in his childhood.

Robert "Bob" Ellsworth moved from New Jersey to Colorado and from Colorado to California, where he took a job as a research chemist with Shell Oil and met there Jack Taylor. Bob became the winemaker at Mayacamas Vineyards in the late 1950s and the general manager in the early 1960s. Eventually, Ellsworth struck out on his own, and in 1965, he started a business, the Compleat Winemaker, renting space in a winery owned by Davis Bynum. The Compleat Winemaker filled an important niche in the area, providing home winemaking supplies to those who lacked the resources or desire to open a commercial winery. Ellsworth's business grew, and in 1968, he relocated to Yountville and began offering specialized commercial winemaking equipment for small producers. Bob would travel back and forth to Europe, purchasing the latest implements in Italy, Germany and France and then bringing them back and modifying them to work with American infrastructure. In this way, Ellsworth helped modernize the wine industry in California and provided smaller operations the opportunity to use the latest and best equipment.

While some focused on making wine, many were content to grow it. One such couple was Tom and Martha May. The Mays purchased a house that came with twelve acres, and the previous owners left them a bottle of wine from nearby Heitz Cellars as a housewarming gift. The Mays were so impressed that they visited the winery to purchase more and began a relationship with Joe Heitz. The Mays eventually purchased more land, planted more vines and named their vineyard after Martha. Soon, Joe Heitz had an exclusive on their fruit, and Heitz Martha's Vineyard became one of the most sought-after wines being produced in Napa.

"The lights at Freemark Abbey came back on in 1967," writes Charles Sullivan in *Napa Wine: A History*. "So dim was the memory of the Ahern and earlier Forni operations that Charles Carpi wondered at first what kind of religious institution had been housed here." A grower named John Bosché had twenty acres in Rutherford that produced premium Cabernet Sauvignon grapes, which he regularly sold to Beaulieu. The year after Freemark Abbey reopened, Bosché began supplying it with grapes, and the wines Freemark produced were quietly heralded as excellent.

By then, Joe Heitz was beginning to scale up production, and he needed to add staff. One early hire was a young investment banker from San Francisco named Bob Travers. Teeming with ambition, Travers had a boxer's chin, a ready smile and far more interest in wine than in banking. Heitz also hired

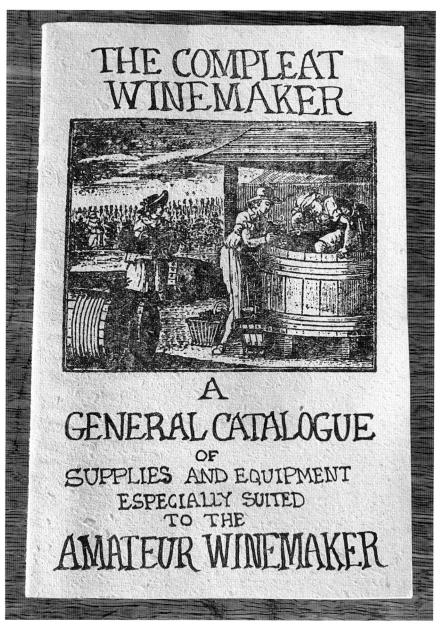

The Compleat Winemaker catalogue. *Photo courtesy Geoff Ellsworth.*

Nils Venge. Venge arrived during a slow period. "Well, Nils," said Heitz, "I'm not going to pay you $2.13 an hour to sit around and look at the wine." Venge was staying in the guesthouse, so his first job at Heitz was to paint it. "I didn't want to just sit around either," smiled Venge with a shrug.

UP IN THE MAYACAMAS and around the same time, Stuart and Charlie Smith were clearing land near the top of Spring Mountain. Ruggedly handsome and physically fit, the brothers poured their own foundation, built their own winery and planted their own vineyards. The land they chose had once been granted by President Chester Arthur to George Cook, and it was covered in many of the same ancient redwoods and madrone trees, which flowered every spring and sprouted reddish-orange berries in the fall, as had been there when Cook assumed the land.

In addition to trees, boulders abounded on the estate, and partway down the mountain, a spring gurgled up from the earth before forming a tiny stream that found its way down the mountain toward the Napa River. Stu and Charlie harnessed the spring and took advantage of vineyard sites that Cook had cleared the century before, and the Smith-Madrone winery emerged into being.

Stu and Charlie Smith in the early days of Smith-Madrone. *Courtesy Smith-Madrone.*

By the mid-to-late 1960s, the Taylors had had enough of living in isolation in a range of massifs named for the mountain lions that still prowled beneath the evergreens, so they sold their shares of the historic winery to Bob Travers, who, after spending a year with Joe Heitz, felt ready to go solo. Bob committed the winery to two varietals, Chardonnay and Cabernet Sauvignon, which he felt sure would do well on the elevated peaks that overlooked the valley floor.

South of the Bay, Paul Draper was busy at work at Ridge. Draper's ready smile inhabited the space between a dignified goatee and a gently receding hairline in a symmetrical head that contained a powerful mind. After the

Bob Travers's handwritten notes about his Mayacamas Cabernet Sauvignon 1971. *Photo by author.*

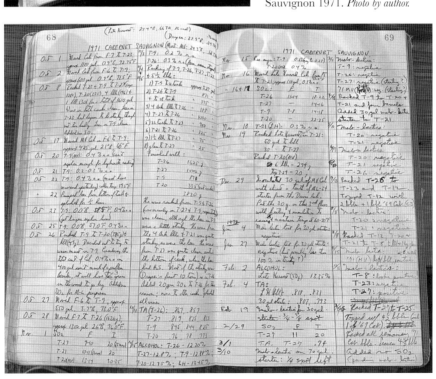

army exposed him to wine in Italy, Paul returned to the United States and worked a short time, as so many had, for Lee Stewart at Chateau Souverain. When the Peace Corps took him to Chile, Draper leased a winery with a fellow volunteer and made wine before traveling to Bordeaux to further his education in the trade. Like Winiarski, Draper didn't have the chance to study winemaking formally, but equally like Winiarski, he didn't let that prevent him from learning to make extraordinary wine. He accepted the job as winemaker at Ridge and almost immediately began to produce elegant wines from this historic mountaintop winery with a view of San Francisco in the distance.

AL BAXTER WAS BORED again. This happened from time to time. In the past, when bored, Al would take up a new hobby. That's how he became a mountaineer. It's also how he started making wine in his basement. A doctorate from Berkeley in philosophy was evidence enough of Al's abilities, but he needed an outlet, or several of them, to occupy his time. When that short Russian winemaker invited himself over to try Al's wines, he was slightly surprised, but what wasn't surprising to Al was that Tchelistcheff enjoyed the wines—so much so that he encouraged Al to make them commercially. Opening a winery seemed about as good a cure for his boredom as any other Al could think of, so he partnered with some investors, including Ron Fenolio, and in 1972 opened Veedercrest Vineyards.

That same year, a more established Napa winemaker, Bob Trinchero, made a blush wine out of Zinfandel grapes. It wasn't a completely original idea, nor was it being done in the Napa Valley much at that time. Trinchero's blush was aged in oak and made in a dry style at first. A few years later, however, a stuck fermentation would lead to elevated levels of residual sugar, rendering it significantly sweeter, still just as flavorful, still an attractive near-neon pink. The popularity of this wine, dubbed White Zinfandel, skyrocketed, and it didn't take long for sales of this product to far surpass those of Bob's red wines, doubling them by the early 1980s. In a nation still nursing its hangover from the sweet home vintages of Prohibition, "White Zin" became a gateway drug for many into the world of American wine.

In still more ways, 1972 was a year of destiny for the Napa Valley. That year, President Nixon raised his glass in a "Toast to Peace" with Chinese premier Zhou Enlai. Nixon hosted the event in Beijing, and both nations committed to working together to build a peaceful future for the world. In his remarks, President Nixon concluded by saying:

Chairman Mao has written: "So many deeds cry out to be done, and always urgently. The world rolls on. Time passes. Ten thousand years are too long. Seize the day." Seize the hour. This is the hour. This is the day for me, two peoples to rise to the heights of greatness which can build a new and a better world.

And in that spirit I ask all of you present to join me in raising your glasses to Chairman Mao, to Prime Minister Chou, and to the friendship of the Chinese and American people, which can lead to friendship and peace for all people in the world.

When he raised his glass, President Nixon held aloft a 1969 Blanc de Blancs produced by Jack Davies at his newly reborn Schramsberg Winery. It was the first time an American wine had ever been served at a White House event. It wouldn't be the last.

At Freemark Abbey, winemaker Jerry Luper was allowing his time in France to influence his decisions—and it was working. His Chardonnays were elegant, balanced and crisp. With his Cabernet Sauvignon, he was blending Merlot to offer greater balance, more complex flavor profiles and a touch of velvet to the mouthfeel. Luper's wines were winning best of show with a high degree of frequency, and Josephine Tychson's former winery was on the rise again.

ANDRÉ TCHELISTCHEFF HAD ENDURED enough nonsense. His beloved Beaulieu—the winery that he and the de Latours had worked so hard to rebuild—had been sold to Heublein, a soulless East Coast corporation that had also purchased Inglenook and systematically sacrificed quality in the name of profit. Innately loyal to the soil and the vines, Tchelistcheff had taken orders from suits whose manicured fingernails would never deign to scrape the topsoil for as long as he could tolerate; he loved Beaulieu, but after being berated for an interview he gave to the *San Francisco Chronicle* and subsequently told that he could no longer speak publicly without a Heublein flunky present, he resigned on April 1, 1973. On April 2, he got in his truck and drove to nearby Sonoma County to begin work as a consultant at the winery established by Giuseppe and Pietro Simi a century before. By then, the Simi Winery had been purchased by Russ Green, another oil executive, and the opportunity to work for a family again—rather than a corporation—appealed greatly to Tchelistcheff. A new chapter had begun.

At Chateau Montelena, Mike Grgich and his team took full advantage of the excellent year and set to work making their second vintage of Chardonnay. Pleased with the first vintage yet convinced he knew how to make a great wine greater still, Grgich was enthused about the possibility of imparting subtler oak-born characteristics into the wine by using the same 225-liter French oak barrels they had used the year before. Most of the grapes were sourced from Alexander Valley, not far from Simi. Father Vincent Barrett, Jim's brother, came from Los Angeles to bless the grapes as they arrived at Chateau Montelena. Grgich felt strongly that this would be a special wine.

Twenty miles down valley, the Trefethen family had purchased the old Eschol Winery. The once-proud redwood facility designed by Hamden McIntyre had, like so many around it, fallen into disrepair and was inhabited by a spectrum of birds, bats and varmints. Eugene and Catherine Trefethen saw promise—and their future—in the historic structure and set about repairing it, pouring a new foundation and partnering with a French producer who had taken an early interest in the Napa Valley. By 1973, harvest was taking place again at the winery, now known as Trefethen. The Trefethens also produced the wine for the Champagne house Moët & Chandon while nearby Domaine Chandon was being built.

At the same time, east of Trefethen, Warren and Barbara Winiarski were joyfully harvesting the fruit from their vines for the first time ever. Planted only three years prior, the young vines had thrived in the shadow of the Stag's Leap outcroppings, and though conventional wisdom might have steered the Winiarskis away from crushing fruit so young, the couple were eager to progress to the next stage of their ongoing reverie. Warren had, by then, acquired a formidable skill set from his time working with Lee Stewart, Robert Mondavi and others—not to mention having spent the past decade reading everything written about wine that he could get his hands on—and he committed himself to pouring all of his knowledge and experience into the best wine he could make.

On the opposite side of the valley from Winiarski, Mike Robbins, an Iowan transplant to San Francisco by way of three tours in Korea, was producing wine of his own. Robbins had briefly been partnered with the Taylors in Mayacamas but sold his shares to seek his fortunes elsewhere. Working in real estate gave Robbins an edge when looking for land, and he began investing in the Napa Valley, buying up hundreds of acres and old structures whenever he could smell out a good deal. He grew Chardonnay and Cabernet Sauvignon and had a winery he called Spring Mountain. In 1974, Robbins was busily buying up more land and beginning the

Warren Winiarski atop a tractor at Stag's Leap Wine Cellars. *Courtesy Warren Winiarski.*

restoration of a mansion, Miravalle, that he had purchased. The house was gorgeous but in disrepair—it looked as though someone had been roller-skating on the hardwood floors—but it was beautiful, nevertheless. Robbins split his time between real estate and viticulture, eventually hiring Charles Ortman, who had previously worked for Joe Heitz, to serve as his full-time winemaker.

Elsewhere, David Bruce, a dermatologist who had taken an interest in wine in college and become a hobby winemaker, was crafting wines on a small plot of land near Los Gatos in the Santa Cruz Mountains near Ridge and Martin Ray. Bruce wanted to try to create an American wine on par with the Richebourg that had sucked him headfirst into wine. Critics began to take notice of Bruce's Chardonnays, and Bruce joined Martin Ray and Paul Draper in helping reestablish viticulture south of the bay.

FARTHER SOUTH, DICK GRAFF had taken note of the climates and soils at Chalone so reminiscent of those in Bourgogne. Limestone and clay, or marl—which is, in essence, a combination of the two—along with scatterings of gravel and sand were home to grapevines that benefited from the cooling presence of the nearby Pacific. Here, Frenchman Charles Tamm had identified that the marine deposits found in the Gavilan benchland north of the Salinas River were quite similar to the Jurassic soils found in his native Burgundy and had planted his grapevines just before Prohibition did its best to choke the life out of the United States.

Graff came to winemaking by way of the usual meandering route taken by so many of his contemporaries. He studied music at an Ivy League university, joined the navy and then become a banker. The natural next step was, of course, to make wine. Graff spent a brief time at UC Davis, enough to learn the basics, then returned to Chalone and got to work.

He began by making wine in an old chicken coop, but by the early 1970s, he had done well enough to build a winery. Graff, the jack-of-all-trades who couldn't resist studying anything and everything around him, plumbed the new winery himself. His wines drew André Tchelistcheff south of the bay to taste them, and his response was so positive that soon, Julia Child and others also took note. Graff's was a small operation, and his wines became widely sought, elevating the price per bottle to well over what most of his competitors were charging.

In 1975, the Napa Valley Grapegrowers Association was formed, and John Trefethen was named president. That same year, the Hollywood film director Francis Ford Coppola, having recently directed the sequel to *The Godfather*, purchased part of the estate that once belonged to the Finnish ship captain Gustave Niebaum, also owned in part by the Heublein people whom Tchelistcheff found so indefatigably irksome. Nearby, Caymus Vineyards, a nod to George Yount's Rancho Caymus, was opened by the Wagner family, natives of the Napa Valley, who hired Randy Dunn, a doctoral candidate at the University of California, as their winemaker.

Enthusiasm for what could be done in the excellent climate and diverse soil compositions that lay between the Vaca and Mayacamas ranges was at an all-time high for the post-Prohibition era and generated enough speculative energy that sixty-some wineries were either born or reborn. Nils Venge purchased land in Oakville with his father-in-law in the hopes that they, too, might open a winery. The Napa Valley was abuzz with the romantic notion of wine and the delectable yet unlikely aspiration of becoming a world-class wine-growing region. A renaissance was certainly taking place, though in the early months of 1976, it was mostly the locals who were aware of it. One exception was Sir Peter Michael.

Sir Peter Michael was an engineer who had worked for Rolls-Royce before starting a business of his own. In the early months of 1976, he found himself in the Napa Valley, not for wine but for love. "I had a passionate affair with Peggy Lee, the singer," recalled Michael. "Of course, I'd never met her, and she didn't know about this—it was quite one-sided, really," he laughed. Peggy Lee was scheduled to perform in Napa, and that was enough to get the Englishman to the valley. At a restaurant, he ordered a bottle of Bordeaux to have with the meal. "It was terrible," he recalled, "So I asked the somm what the locals drink." The sommelier returned with a bottle of Chateau Montelena. "It blew our socks off, really!" he said. "I suddenly realized that California wine didn't come in a box with a tap on the front."

Only his wife, Bella Lawson, could have said for sure if Steven Spurrier did not wake up every morning with his hair in place, wearing a suit jacket and a collared shirt. A gentle, soft-spoken Englishman with bright eyes and a warm smile, Spurrier was the proud owner of a small wine shop in Paris. He was prim, trim and always impeccably dressed; his kind voice was as disarming as his palate was impressive; and his enthusiasm for wine was rivaled only by his knowledge.

Spurrier had come to own his wine shop by promising its recently widowed Parisian owner—who was understandably somewhat reluctant to sell her husband's pride and joy to, of all things, an Englishman—that he would work there for free for six months to demonstrate his commitment. He did, and in six months' time, the shop was his. Soon after, Spurrier opened the Académie du Vin in partnership with Jon Winroth, an American Fulbright Scholar turned wine writer from the *Herald Tribune*. They then hired another Parisian outsider, an American by the name of Patricia Gastaud-Gallagher. English speakers peddling France's most prestigious and elite cultural wares back to the French—why not? However cockamamie the idea might have sounded, it worked; Spurrier and his associates knew that their audience was not, in fact, the French and took out ads targeting expats who found English-speaking wine merchants less intimidating and, thus, easier to buy from.

Gallagher began to teach at the Académie du Vin, and through friends at the U.S. embassy, she annually arranged a tasting of American wines on July Fourth. What few American wines had any share of the French import market at that time, however, hardly constituted cream having risen to the top. Looking to increase the quality of the American wines she offered, Gallagher telephoned a woman by the name of Joanne Dickenson DePuy who had lived in Napa since 1951 and who was aspiring to start a business offering international wine tours. The women spoke a while, and DePuy suggested some of her personal favorites to Gallagher.

Anyone who might have been aware of this conversation would surely have found it innocuous, even mundane. And yet, in the same way that the golden apple of Eris would result in the fall of Troy, DePuy's words to Gallagher that day would, in due course, prove to be a golden apple of her own, bringing the walls surrounding the myth of superior French wine crashing to the ground.

Chapter 3

THE TROJAN HORSE

No man or woman born, coward or brave, can shun his destiny.
—The Iliad

Steven Spurrier was as audacious inwardly as he was charming outwardly. He grew up in Marston Hall, the family home in Derbyshire, England, the son of landed gentry fifty years too late for that to matter much, and the world had rested at his delicate fingertips since the advent of his charmed if exceedingly labyrinthine life. As a boy, Spurrier entertained himself by playing in the estate's wine cellar, and as a young man, he took to extensive traveling in France, Spain, Germany and Portugal, all in the name of studying wine. From an early age, there was little doubt that the beguiling and charismatic young man was destined for a career in wine.

At the age of thirteen, the intelligent and curious Spurrier had his first taste of wine. "My grandfather said that he thought I was old enough to try the port and asked the butler to bring me a glass," recalled Spurrier. The port on the Spurrier family table that Christmas Eve was Cockburn's 1908. "The wine…was quite extraordinary, and the impression it left has lasted a lifetime," Spurrier shared. "That was my Damascene moment, the moment when the seed was firmly planted for my life in wine."

Spurrier met his wife, Bella, one evening while working at Christopher's, the oldest wine merchant in London. Bella Lawson was slender, attractive and smart as a whip. The young Steven had reluctantly taken off his ice skates, which returned him to his normal height, and joined a friend in the bar near the Queensway Ice Skating Rink in Bayswater. The young Steven found girls intimidating, and he let his friend Ian Carr do most of the talking. "This is my friend Steven, who'd like to offer you a drink," was Ian's contribution. The girls took him up on the offer, but while the first two ordered cider, the third, seventeen-year-old Bella Lawson, asked for a gin and orange.

Steven's mother had once quipped within earshot that gin and orange was the preferred drink of prostitutes, and remembering this, the bashful Steven somehow summoned the courage to say to Bella, "You can't possibly— that's what tarts drink!" Undaunted, Bella responded, "In that case, make it a double." Later in life, Bella would laugh as she recalled their meeting. "I don't think I'd ever drunk gin," she admitted. "I thought it sounded a bit sophisticated, I expect." However it sounded, Steven was smitten. "We got on right from the start and still do," he later wrote in his memoir. The pair were joined in matrimony in 1968 and enjoyed more than fifty years of marriage.

Steven and Bella Spurrier eventually made their way to Paris, where they lived on a barge moored on the Seine. Steven courted the business of British and American expats in Paris, people who—like him—spoke middling French and who would prefer to avoid the embarrassment of mispronouncing words while they shopped for their wine. Spurrier also sought to appeal to Paris's elite French wine establishment. According to George Taber, Spurrier courted these elite palates "with all the subtlety of a bulldozer," making a point of attending every important wine-related event in the city and introducing himself as the owner of the Caves de la Madeleine—whether he had been invited or not.

Spurrier "set the standard for elegance in the French retail world" and was the first *caviste* to wear a necktie daily. He made a point of only selling wines from producers he knew personally, and his dedication to small, family-owned wineries was noteworthy. Furthermore, Spurrier was not a salesman and had no intention of becoming one. Rather, he was an educator, and his approach to the sale of wine was to teach people about what was in the bottle and let them decide for themselves. It worked. Remarked Jancis Robinson, arguably the most influential wine writer of the age: "If he has a fault, it is hardly the most serious: an excess of enthusiasm about even the most humdrum of wines."

At the same time that Spurrier was working hard to get his wine shop going, Patricia Gastaud-Gallagher was enjoying living in Paris and had no interest in returning permanently to her home in the United States. There was just one problem with her desire to remain indefinitely on holiday in France: she had no way to pay for it. She was a talented writer with a great mind, but that and four francs might buy her a glass of Cru Bourgeois in most cafés in Paris. Gallagher needed to find a steady job if she wished to continue living in the City of Light. Gallagher didn't know much about fermented grape juice, but Spurrier admired her intellect and, given that there weren't a ton of people beating down the door for a job at his experiment, he hired Gallagher to work at the Académie du Vin.

Spurrier soon earned an excellent reputation. When Queen Elizabeth II hosted a dinner at the British embassy for French president Georges Pompidou, her staff went to Spurrier for wine. In league with a fellow Briton who was growing grapes in southern England, Spurrier got the idea to serve English wine to the president of France at the dinner. He ordered five cases of dry white wine and got the dinner planners to add it to the menu. Soon, however, Spurrier got a call from French customs agents. The wine would be shipped back to England, they informed him. When an outraged Spurrier asked why, he was informed by the agent that "English wine does not exist."

The agent explained to Spurrier that the lengthy list of things that could be imported from England did not include wine, at which point the usually kindly Spurrier explained to the agent in blunt, cudgeling terms that the wine was intended to be served that very night to President Pompidou and Queen Elizabeth II, and that if the wine did not exist, the customs agent's job would soon be equally nonexistent. That evening, Queen Elizabeth II served English wine to President Pompidou and their esteemed dinner guests.

After her fateful conversation with Joanne Dickenson DePuy, Patricia Gastaud-Gallagher proceeded to make a trip to the Napa Valley in the autumn of 1975, returning soon after to France and gushing to Spurrier about the quality of wine being produced in Northern California. When Gallagher pitched the idea of tasting American wines in France on the bicentennial of American independence, Spurrier responded, "That's not something we Brits normally celebrate, but I'll go along with it."

Gallagher's impressions were strong enough to prompt Steven and Bella Spurrier to embark on their own excursion to learn more about Californian wines and the people making them. It wasn't necessarily easy. Joe Heitz, whom Steven later described as a "charming curmudgeon," told him bluntly, "If you are a journalist, I don't receive them; if you are a merchant, I don't export; and anyway I don't have time." Ridge, similarly, gave Spurrier the brush-off. Spurrier won Heitz over by sincerely suggesting that Joe's Chardonnay reminded him of a Meursault, Alice Heitz's favorite wine and thus the bar Joe was trying to clear. At Ridge, the Spurriers simply drove two hours south of Napa, up the switchbacks to Monte Bello, and knocked on the door. "Aren't you the fella I told not to come?" asked David Bennion, one of the owners, adding, "Well, since you're here, you'd better taste some wine!"

To ensure things went more smoothly at the other wineries, the Spurriers enlisted the help of Joanne Dickenson DePuy. DePuy enjoyed showing the Spurriers around, and her impressions of Steven and Bella were much the same as those of everyone else who encountered them: the Spurriers were warm, polite and charming. As the Spurriers prepared to return to France, DePuy was preparing her own trip to the same location. André Tchelistcheff had, after much pestering, agreed to lead a tour of France's most esteemed wine regions on behalf of DePuy. Less than a week prior to their departure, DePuy received an urgent call from Steven Spurrier.

If getting supposedly fictitious English wine into France for the Queen had been difficult, it was nothing compared to importing equally mythical American wines for seemingly no good reason. Spurrier and Gallagher had meticulously selected boutique wines. They had arranged for the wine tasting to take place at the historic Hotel InterContinental in Paris. They had organized a panel of exceptionally well-credentialed tasters. They were prepared in every way except one: they didn't have the American wines in Paris, and they had no way of getting them there. Spurrier realized too late that he could not personally transport that much wine through customs—in fact, he was allowed only two bottles and Bella another two, leaving them in desperate need of assistance.

Robert and Margrit Mondavi.
Courtesy Napa County Historical Society.

DePuy called around and was advised that, like Spurrier or any other individual with a ticket, every member of her tour group could account for two bottles at French customs. Among those in the group were André with his wife and their grandson, Paul, accompanied by Napa royalty that included Jim and Laura Barrett; Andy and Betty Beckstoffer; Margrit Biever, who would soon become Margrit Mondavi by marrying Robert; MaryAnn Graf; Zelma Long; Louis P. and Liz Martini; Bob and Nonie Travers; and more. Zelma Long

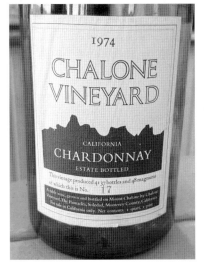

Top, left: A bottle of the 1973 Chateau Montelena Chardonnay, made by Mike Grgich, that won the white wine category of the Paris tasting. *Courtesy Chateau Montelena.*

Top, right: David Bruce Chardonnay 1973 (late harvest). *Photo by author.*

Bottom, left: Chalone Chardonnay 1974. *Photo by author.*

Bottom, right: A bottle of the 1973 Stag's Leap Wine Cellars Cabernet Sauvignon, made by Warren Winiarski, that won the red wine category of the Paris tasting. *Courtesy of Stag's Leap Wine Cellars.*

Top, left: Ridge Cabernet Sauvignon 1971. *Photo by author.*

Top, right: Freemark Abbey Cabernet Sauvignon 1969. *Photo by author.*

Bottom, left: Clos du Val Cabernet Sauvignon 1972. *Photo by author.*

Bottom, right: Mayacamas Cabernet Sauvignon 1971. *Photo by author.*

had been invited by Martini. "It was one of those 'Wheeeeee!' wonderful experiences," she later recalled.

Assembled, the group became a Trojan horse, and divided among the travelers, the soldier-bottles stole into the country virtually undetected. On May 7, 1976, the tour deplaned at Charles de Gaulle Airport with Spurrier's precious cargo. Dressed impeccably in a white summer suit, Spurrier met the group at the airport, treated them to a light lunch, thanked DePuy, shook hands with Tchelistcheff and set off into Paris with the wines in tow.

That transporting wine that supposedly didn't exist into the country that considered itself the alpha and omega of viticulture wasn't easy was not the only thing Spurrier realized as he prepared the tasting. He also knew that tasting was highly subjective and that, as much as he respected them, his French judges would be biased. "The risk was they'd know that California was on the West Coast somewhere north of Mexico and they would think it was a southern country and they'd judge it like the south of Spain or the south of Italy," recalled Spurrier. Paul Draper, the winemaker at Ridge, was more direct. "[Spurrier] began to realize that, the French being the French, [they] would simply say when they tasted them, 'Ah, America is so interesting,' and all that, and that would be that." The more Spurrier thought about having his highly credentialed panel taste the American wines, the clearer it became that he needed to build in a safety mechanism.

THE JUDGES, A VERITABLE who's who of the French wine trade, arrived at the Hotel InterContinental in Paris a little before three o'clock in the afternoon. This Spurrier fellow was an odd bloke, to steal a term from his fellow Englishmen, but likeable enough. He knew their wines, producers and regions better than any of them—though they'd never have said so—and he seemed a Francophile at heart. True, the notion of tasting American wines was rather silly, perhaps, but Spurrier had assured them of the wines' quality and of his unending gratitude for their involvement. They would show up, say hi to Steven, taste some wine and enjoy an evening meal on the Seine. Nothing so wrong about that.

The InterContinental, in the heart of Paris, provided a perfect space. This was due in part to Spurrier's relationship with the hotel's food and beverage manager, Ernst Van Dam. The well-lit room was on the ground floor, with large picture windows leading out into the courtyard. The only catch was that they had to be out before six o'clock in the evening so the hotel could set up for a wedding.

The lone journalist to attend was a tall, young, good-looking American named George Taber, who worked for *TIME* magazine in Paris and spoke perfect French. Gallagher had remembered that Taber took a wine course from her and extended the invitation. "He accepted, with the caveat that if something else cropped up, he wouldn't be able to make it," Spurrier recalled. Fortunately, nothing did.

As the judges arrived one by one, Spurrier and Gallagher greeted them enthusiastically, thanking them again for taking part. As Taber would later recount, "The nine French judges…included such high priests as Pierre Tari, secretary-general of the Association des Grands Crus Classés, and Raymond Oliver, owner of Le Grand Vefour restaurant and doyen of French culinary writers."

The other seven judges—though Spurrier confessed that even he did not consider them to be judges but rather tasters, at the time—were indeed a formidable ensemble. The impeccably proper Odette Kahn gave off a palpable air of strength and confidence. One of the few powerful women in a field then—and still—dominated by men, Kahn was the editor of two important wine publications and was as strong as she was direct.

Pierre Brejoux was the inspector general of the Appellation d'Origine Contrôlée Board and had authored numerous books on wine—French wine, of course. Christian Vannequé, age twenty-five, was the head sommelier at the most famous restaurant in Paris, La Tour d'Argent. Aubert de Villaine was the co-owner of Bourgogne's famous Domaine de la Romanée-Conti, the wines from which today can easily fetch upward of $25,000 a bottle—if, indeed, a person can locate them at all. Jean-Claude Vrinat had worked in his father's restaurant, Taillevent, as it rose to fame after the Second World War. In 1973, the restaurant earned its third Michelin star, and Jean-Claude became general manager.

Among this who's who, there were also two "who the hell?"s. Michel Dovaz was a knowledgeable unknown who taught a course at the Académie du Vin. In the hopes of better fitting in with this prestigious gaggle of fellow judges, Dovaz told people he was the president of the Institute Oenologique de France, though it seems terribly unlikely that any of them would have been convinced, as such a thing did not exist. The other unfamiliar face on the panel belonged to Claude Dubois-Millot, who was, in fact, attending his first wine tasting. Dubois-Millot had spent decades working in the automotive industry and recently taken a position at the French publication *GaultMillau* as sales director. The publication had at first agreed to cover the tasting, thought better of it and sent the inexperienced Claude as a gesture of goodwill to Spurrier.

When all the judges had, at last, arrived, Spurrier showed his hand: "I've decided to change the rules," he said plainly, then offered, "If you don't agree, I'll change them back." In fact, Gallagher herself initially disliked the idea. "I did it anyway," recalled Spurrier.

"Instead of just tasting Chardonnays and Cabernets from California, I thought it would be interesting to match them against some benchmarks of similar vintages from Burgundy and Bordeaux," he told the judges, terming the idea "a friendly 'glasses across the ocean' blind tasting" in celebration of American independence and in acknowledgment of the important role that the French general Lafayette played in the American Revolution. Spurrier recalled murmurs of *"Bonne idée"* and *"Pas de probleme,"* with no objections whatsoever. Spurrier then handed out the tasting sheets, and Gallagher handed Taber a copy of the tasting order so that he alone would know which wines were being tasted.

According to Taber, the tasting began with a glass of Chablis—Chardonnay produced in the northernmost part of Bourgogne—to awaken the palates of the judges. From there, the white wines were tasted first. All told, there were six American wines and four French in the white category. The American wines were all Chardonnay and included Napa's Chateau Montelena 1973, Spring Mountain Vineyards 1973, Freemark Abbey 1972 and Veedercrest 1972, along with David Bruce 1973 and Chalone 1974 from farther south.

While these American producers were entirely unknown in France, the French white wines were all renowned and of the highest pedigree, hand-selected by Spurrier. "I certainly did not stint on the origin and quality of the French wines to give the California wines a head start!" Spurrier later remarked. The French whites included a Puligny-Montrachet 1972 and three wines from the 1973 vintage: Meursault Charmes, Beaune Clos des Mouches and Batard-Montrachet.

As the tasting ensued, Taber observed that the judges appeared nervous and uncertain—hardly the posture one would expect from the most respected wine experts in all of France. "There was lots of laughing and quick side comments," Taber wrote. Taber had the run of the place due to other journalists' and news outlets' refusal to take the tasting seriously and soon realized that the judges weren't only nervous, they were also confused. They talked more than was customary and often disagreed with one another about the origin of a wine. Taber took note of a number of instances in which a judge openly criticized a wine and, in the same breath, declared it American, when in fact the wine was French. Similarly, judges would compliment a wine's quality, certain of its French origins, and Taber

Above: *Left to right*: Gallagher, Spurrier and Kahn. One of seven photos of the Judgment of Paris taken by Bella Spurrier. *Courtesy Bella Spurrier.*

Left: The judges assembled. One of seven photos of the Judgment of Paris taken by Bella Spurrier. *Courtesy Bella Spurrier.*

Above: The judgment begins. One of seven photos of the Judgment of Paris taken by Bella Spurrier. *Courtesy Bella Spurrier.*

Left: Taking notes. One of seven photos of the Judgment of Paris taken by Bella Spurrier. *Courtesy Bella Spurrier.*

Top, left: Aubert de Vilaine, nose-deep in Chardonnay. One of seven photos of the Judgment of Paris taken by Bella Spurrier. Courtesy Bella Spurrier.

Top, right: Revisiting the Chardonnay. Steven Spurrier is in the foreground. One of seven photos of the Judgment of Paris taken by Bella Spurrier. *Courtesy Bella Spurrier.*

Bottom: Confusion sets in. One of seven photos of the Judgment of Paris taken by Bella Spurrier. *Courtesy Bella Spurrier.*

would cross-reference his cheat sheet before silently jotting a note to himself: the wine was from California. After one judge tasted the Freemark Abbey Chardonnay and responded aloud, "Ah, back to France," Taber thought to himself: *Hey, maybe I'm going to have a story here after all.* As the white wine tasting concluded, Taber spoke to one of the judges, who openly admitted: "Our confusion showed how good California wines have become."

In a hurry to get out of the InterContinental in time for the staff to set up for the wedding, Spurrier announced the results of the white tasting while the reds were being set up. The judges were horrified. They themselves had ranked an American wine first—and by a wide margin, no less? They themselves had awarded California wines three out of the top five places? How was this possible? The one wine they had been able to identify as American, the David Bruce, they had been sure to properly disparage, but beyond that, the quality of the American wines clearly matched and even exceeded that of the French producers'. The results, as read aloud by Spurrier to the consternated panel, were as follows:

First: Chateau Montelena 1973 (California)
Second: Meursault Charmes 1973 (France)
Third: Chalone 1974 (California)
Fourth: Spring Mountain 1973 (California)
Fifth: Beaune Clos des Mouches 1973 (France)
Sixth: Freemark Abbey 1972 (California)
Seventh: Batard-Montrachet 1973 (France)
Eighth: Puligny-Montrachet 1972 (France)
Ninth: Veedercrest 1972 (California)
Tenth: David Bruce 1973 (California)

Taber admits to having been uncertain who won, given the French word *château* in the name of the Napa Valley winner, but Gallagher quietly dispelled his doubts. According to his own account, Taber felt a sense of pride that, as a journalist, he did not vocalize in the moment.

Given the results, it is somewhat surprising that the judges even proceeded to taste the red wines at all, yet indeed they did. This time, according to Taber, "the judges seemed both more intense and more circumspect." Taber continued to monitor their comments and compare their remarks to his list. "Their comments about the nationality of the wine in their glasses were now usually correct," wrote Taber. The increased focus and accuracy of the panel, however, could do nothing to detract from the quality of the wines they were tasting. After they had concluded their work and Spurrier had tallied the results, he again read them aloud to the room:

First: Stag's Leap Wine Cellars 1973 (California)
Second: Château Mouton Rothschild 1970 (France)
Third: Château Montrose 1970 (France)
Fourth: Château Haut-Brion 1970 (France)
Fifth: Ridge Monte Bello 1971 (California)
Sixth: Château Léoville Las Cases 1971 (France)
Seventh: Heitz Martha's Vineyard 1970 (California)
Eighth: Clos du Val 1972 (California)
Ninth: Mayacamas 1971 (California)
Tenth: Freemark Abbey 1969 (California)

According to Taber's account, "This time the stir in the room was even more pronounced than before." Stunned himself, the journalist again consulted Gallagher to confirm that he was correct, that an American wine had triumphed a second time that afternoon. "Yes," replied the composed Gallagher simply.

Others in the room were less composed. Odette Kahn, who was in publishing on the wine side, was probably the best positioned to see through the annals of time and predict the sort of impact this could have on the industry. Though Taber never remarks on it in his writing, his presence as an American journalist for *TIME* magazine, standing there scribbling furiously in his reporter's notebook, no doubt made her deeply uncomfortable. Kahn, of course, would have known better than to attempt to silence the press, so instead she went after Spurrier, demanding to have her scorecards back. "You're not going to get them back," Spurrier informed her resolutely.

None of the other judges were as aggressive, nor seemingly as concerned, as Kahn. After the tasting, they lingered in the InterContinental, sipping Champagne and remarking on the tasting. Many spoke to Taber, and of those, all seemed to agree that while the California wines had undoubtedly shown well that day, the French wines still, in their professional opinions, remained superior. But they had already registered their objective opinions on scorecards, which by late afternoon were secured back at the Académie du Vin. The time for their opinions had passed. It was the opinion of the wine-buying public that mattered now.

WARREN WINIARSKI WAS IN Chicago visiting family when he received a call from Barbara, back in Napa. "Dorothy Tchelistcheff just called!" blurted Barbara. "Do you remember the tasting? The one in Paris?" Warren wracked his memory. Paris? "You know, the one the English gentleman set up?" Barbara persisted. "With your Cabernet?" Warren didn't really remember, but he could tell that this was important to his wife, so he played along. "Er, sure," he told her. "Well, we *won*!" cried Barbara. "Dorothy said we won the tasting!" *How do you win a wine tasting?* wondered Warren. "That's nice," he told his wife before hanging up.

Jim and Laura Barrett had eagerly joined Tchelistcheff's tour de France and had been part of the expedition that transported the wine over the ocean on Spurrier's behalf. They were just sitting down to lunch at a winery in Bordeaux when someone from the estate approached Jim and informed him that he had a phone call. It was George Taber, looking for a quote to go along with his article for *TIME*. Never one to boast and realizing that whatever he said had the potential to follow him forever, Barrett downplayed his enthusiasm for what the journalist had just shared with him and crafted a quick colloquialism for Taber that was part axiom, part proverb.

"Not bad for kids from the sticks," said Barrett, smiling to himself and eager to hug his son, Bo. Jim returned to the table and leaned in close to whisper into Laura's ear. "Our wine won in Paris," he told her. He then sent a telegram back to Calistoga, informing Bo, Mike, a newly hired Gustavo Brambila and the rest of the staff of their victory. Grgich was beside himself with joy and began to do a Croatian jig around the barrel room. "Pretty soon we were all hopping around, learning the Croatian jig," Bo Barrett later recalled.

It was only once the tour of Americans had climbed back on the bus, having endured a condescending, patronizing speech from their unwitting Bordelais host about how, if they continued to work very hard, someday the Americans, too, might produce some very good wines, that Barrett, Tchelistcheff and the rest of the tour allowed themselves to express their exuberance. There was cheering, hugging and a permeating sense of awe and optimism enveloping everyone within the charter, filling them with a sense of excitement and eagerness about the future. "Hot damn!" exclaimed Barrett. "We knocked 'em in the creek!"

Two weeks later, on June 7, Taber's article appeared in *TIME*, buried near the back, titled "Judgment of Paris." Part journalist, part poet, part scholar, Taber was riffing on Greek mythology. Eris, goddess of discord, was not invited to Achilles's parents' wedding. In response to the slight, Eris

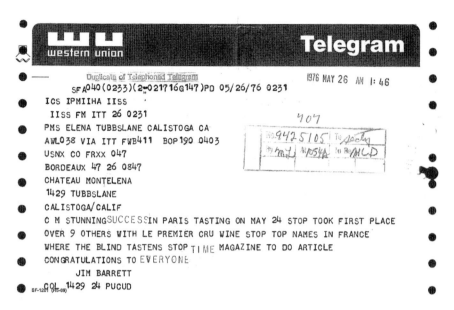

The telegram from Jim Barrett to Chateau Montelena informing his staff of the victory, dated May 26, 1976. *Courtesy Grgich Hills.*

tossed a golden apple, on which "to the fairest one" was inscribed, amid the guests. Athena, Hera and Aphrodite squabbled over the prize. Too wise to judge the fairest himself, Zeus tasked Prince Paris of Troy with doing so. By judging Aphrodite the fairest, Paris was granted the love of Helen and the ire of Athena, Hera and the Greeks. The result of the first Judgment of Paris was the fall of Troy. The result of the second Judgment of Paris was the fall of the myth of indomitable French terroir. Taber's article begins with the line "Americans abroad have been boasting for years about California wines, only to be greeted in most cases by polite disbelief—or worse." He concludes the one-page article with Jim Barrett's modest assessment of the victory: "Not bad for kids from the sticks." Taber wrote seven pages. *TIME* published four paragraphs on page 58. It proved to be enough.

Sir Peter Michael was among the legions who read the story with great interest. "It completely shook the place up," Michael recalled. "I bought the idea of California," he said, and soon after, he bought land there as well, in Knight's Valley, less than a ten-minute drive from Chateau Montelena.

Four days after "Judgment of Paris" was published, the Friday edition of the *Napa Register* ran the headline "French Judge Napa Wines Best" and featured a large image of Warren Winiarski with his nose deep in a glass of red wine. "Napa Valley's finest vintages have again bested leading French

ACTUAL TIME MAGAZINE,
June 7, 1976

Left: TIME magazine cover and article by George Taber, on display at Chateau Montelena. *Photo by author.*

Opposite: The front page of the *Napa Register*, proclaiming victory, dated Friday, June 11, 1976. *Courtesy Napa County Historical Society.*

wines in a 'blind tasting'—but this time, French wine connoisseurs were the judges," touted Steve Hart in the *Register*. Hart went on to list the Napa wines and interviewed Winiarski, who was quick to credit André Tchelistcheff and congratulate Mike Grgich.

The French were not so keen on the story, of course, but neither were they alone in their incredulous resentment of reality. Soon after the *TIME* and *Register* pieces came out, the Sunday edition of the *Washington Post* ran a story titled "Those Winning American Wines," wherein writer William Rice begrudgingly began the story by declaring the tasting "rather pointless," called the Napa wines "expensive" and proceeded to misspell not one but both of the victorious Napa winery's names as "Monthelena" and "Stagg's Leap," respectively.

No publicity is bad publicity, however, and the tasting deemed the Judgment of Paris by Taber—to whom all credit for subsequent coverage is rightfully due—was getting press. "It took time for the news to sink into the wine community consciousness, and longer into the broad wine knowledge,"

French Judge Napa Wines Best

By STEVE HART
Register Staff Writer

Napa Valley's finest vintages have again bested leading French wines in a "blind tasting" — but this time, French wine connoisseurs were the judges.

Wines from Stag's Leap Wine Cellars north of Napa and Chateau Montelena winery near Calistoga were chosen over vintage Chateau Mouton-Rothschild and Meursault-Charmes by the French wine experts.

The Gallic tasters — winemasters at the best French restaurants, European gourmet writers and Continental wine industry officials — were reportedly startled that "ingenue" California vintages would outscore the time-honored French products. "There was general consternation among the judges," reported one local entrant.

The Bicentennial tasting was sponsored by a Paris wine institute, L'Academie du Vin. A half dozen California Chardonnays were pitted against the famous white Burgundies of France, while Cabernet Sauvignon samples from Napa Valley were judged against the classic Chateau red wines from Bordeaux.

Other local products entered were from Freemark Abbey winery, Veedercrest vineyards, Spring Mountain Vineyards, Clos Du Val winery, Mayacamas vineyards and Heitz Cellar.

When a dozen rounds of careful tasting and note-taking were over, a 1973 Napa Valley Cabernet Sauvignon from Stag's Leap Wine Cellars had won the jury's acclaim, beating the French chateau wines from Mouton-Rothschild, Haut-Brion and Montrose.

In the white wine category, the Paris tasters chose the 1973 Chardonnay from Chateau Montelena winery, followed by Meursault-Charmes '73, a distinguished French wine. Chardonnay from Spring Mountain Vineyards in St. Helena finished fourth.

Warren Winiarski, owner-winemaker at Stag's Leap, said the judging "is a landmark for us, and it's good for the valley."

He credited both winning wines to the influence of Andre Tchelistcheff, retired winemaker at Beaulieu Vineyard, considered by many to be California's greatest living vintner.

"I've always admired Andre's wines," said Winiarski. He said that Miljenko "Mike" Grgich, Yugoslavian-born winemaker at Chateau Montelena, was also taught by Tchelistcheff.

Winiarski, a former winemaker at Robert Mondavi Winery and Souverain Cellars, said less than 2000 cases of the prize-winning Cabernet were made. A fifth sells for $7.

One expert taster called the local product a "supple, complex" wine, with "deep garnet color, a fruity aroma, great depth, and a vanilla finish."

The Chateau Montelena '73 Chardonnay is hard to find on retail shelves. "It's sold out," said a winery

(Continued on Page 2)

WARREN WINIARSKI, owner-winemaker at Stag's Leap Wine Cellars, sniffs the bouquet of his award-winning 1973 Napa Valley Cabernet Sauvignon, which bested the classic red wines of Bordeaux in a Paris wine tasting recently. Expert Continental wine buffs chose a Chardonnay from Calistoga's Chateau Montelena winery as the best wine in the whites category, outscoring the premier white Burgundies of France in the "blind tasting." (Register photo by Bob McKenzie)

The Napa Register
and THE NAPA JOURNAL
114th Year No. 260 | Friday, June 11, 1976 | Price: 20 Cents

Hays Recovering, But Pill Mystery Remains

BARNESVILLE, Ohio (UPI) — Rep. Wayne Hays came out of a 20-hour coma caused by an overdose of sleeping pills today and smiled at his wife from a hospital bed. The veteran congressman, target of a Washington sex scandal investiation, was upgraded to satisfactory condition.

Hays, 65, is "saying a few words and making a few short sentences that make sense," said Dr. Richard Phillips, the lawmaker's personal physician. He said Hays asked for his wife, Pat, and "smiled when she came into the room."

Hays, who won the Democratic nomination for his 23rd term in Congress from Ohio Tuesday, was admitted to the Barnsville Medical Center Thursday morning suffering from an overdose of Dalmane sleeping pills. But Phillips said there was no evidence of any suicide attempt.

Hays is the target of two Washington investigations into charges he put Elizabeth Ray on the congressional payroll to serve as his mistress at $14,000 a year.

Phillips said he had removed Hays from the "guarded" list and placed him in satisfactory condition. He said Hayes regained partial consciousness at 5:45 a.m. EDT and asked for his wife.

"His body simply has to detoxify the drug," said Phillips earlier. He added at the midmorning briefing that Hays will have to spend "at least 10 more days in the hospital" and "a week to 15 more days at home" for complete recovery.

"The fact that he is waking up indicates that he did not take a horrific dose" of sleeping pills, said Dr. David Schuster, one of four physicians attending Hays.

Mrs. Carol Clawson, Hays' press secretary, said meanwhile Mrs. Hays had been by her husband's bedside or in an adjoining hospital room since he was admitted. She said Mrs. Hays, who married the congressman in April, had visited her father in Tennessee Wednesday and returned to Ohio Wednesday night.

Phillips said Hays had taken an unknown amount of Dalmane, "a commonly prescribed sleeping pill," apparently late Wednesday night.

No suicide note was found, Phillips said, and Hays had "no suicidal tendencies." He said there is "no direct evidence" that Hays attempted suicide.

Columnist Jack Anderson said today Hays discussed possible suicide with him in Washington last week. He said Hays also telephoned from Ohio Wednesday afternoon to report that his new wife, Pat, had left him twice because of the sex scandal allegations by Elizabeth Ray, a former Capitol Hill office worker.

Anderson said on ABC's "Good Morning America" show that Hays was "deeply despondent" in an hour-long meeting in the columnist's office last week.

"He said in a low, husky whisper that if it would spare her more anguish, he'd put a bullet through his head, (and) he added 'I've got the guts to do it'," said Anderson.

Mrs. Clawson arrived at Hays' bedside at the Barnesville Medical Center about midnight after Hays had been in a coma for more than 12 hours. Phillips and a pair of consulting physicians flown in from Pittsburgh earlier today were scheduled to examine Hays later this morning.

Hays, 65, was found unconscious at his farm home near Belmo Thursday morning by Pat Peak, his wife of two months, and rushed 20 miles by ambulance to the Barnesville Medica Center.

VANDALS caused an estimated $2,000 damage to the American Canyon sewage pumping station Thursday night, smashing chemicals, above photo, and monitoring controls. Another $10,000 loss was suffered by the construction company which is building a $2 million replacement facility. For additional photos, see page 2. (Register Photo by Bob McKenzie)

Lebanon Artillery Battle

BEIRUT, LEBANON (UPI) — A wild artillery battle around Beirut's International Airport shattered a lull in the Lebanese civil war today and leftists charged Syria was sending more infantry, tanks and planes against their men in a bid to impose a Syrian peace on this war-torn nation.

Leftist reports also said Syrian forces in north and south Lebanon heavily shelled their positions in the northern and southern ports of Tripoli and Sidon. Palestinian sources charged Syria was preparing a new offensive.

(A Cairo-based Voice of Palestine radio station reported large-scale fighting throughout Lebanon and said, "The situation has exploded on all fronts.")

A two-day lull in the fighting between Palestinian guerrillas and Syrian controlled Saiqa forces in Beirut was shattered early today when a clash along the airport road erupted into fierce exchanges of artillery and mortar fire.

A Palestine Liberation Organzation spokesman charged the pro-Syrian forces had provoked the attack by shelling the nearby refugee camps of Borj Brajmen and Chatilla, causing "heavy casualties and damages".

Amid reports that Syria was moving fresh troops and additional equipment into the eastern Bekaa Valley, Palestinian sources charged the Syrians were preparing a new offensive against the leftist-Palestinian alliance.

"All oftheir actions indicate they are planning a new assault," one source said. "They have stopped their advance, but they are bringing in reinforcements all over." A Christian radio station said 1,000 trooops of a joint Algerian-Syrian force had already crossed into Lebanon from Syria, but the reports could not be confirmed.

Western newsmen traveling across the border said the Syrians were bringing "dozens" of heavy field guns into the Bekaa but saw no troop movements.

$7 Million Tentative Budget For College

By STEVE TAYLOR
Register Staff Writer

Napa College trustees voted 4-0 Thursday to accept a tentative budget for 1976-77 listing over $7 million in expenditures, $6.6 million in income and a nearly $1 million "ending balance."

The budget itself is expected to see a "deficit" of over $550,700. But Business Manager Andrew Peterson keeps a separate carryover fund which includes money left from previous budget years.

Nearly $1.5 million is expected to be left over from the current fiscal year's financial operations. This, plus expected income, leaves school officials with a fund totalling nearly $8.3 million for the 1976-77 fiscal year.

Peterson expects $941,468 to be left over when the next fiscal year comes to an end.

He said the tentative budget is a "status quo" document and expects the budgeted expenditures to increase prior to final budget adoption.

A public hearing will be held Aug. 5.

If last year's budget process is any indication, the tentative budget will increase considerably. The tentative budget for 1975-76 was about $6.3 million. When the spending plan was finally adopted in August, expenditures were projected to total nearly $7.8 million.

Actually, expenditures for 1975-76 will probably come in closer to $7 million, according to Peterson's "reestimate."

This is the second year the budget will operate at a "deficit," meaning that "income" (not counting the funds carried over from previous years) will exceed expenditures, Peterson noted.

(Continued on Page 2)

Vandals Hit AmCan Sewage Plant; $10,000 Damaged Or Stolen Goods

AMERICAN CANYON — Vandals struck the construction site of a $2 million sewage pumping station Thursday night, causing $5,000 worth of damage and carrying off another $5,000 worth of tools, officials reported this morning.

Construction will be shut down for two to three days while workers re-equip themselves and restore order to the office trailers which were made a shambles, said Larry Jenkins, supervising engineer for Peter Kiewit Sons Co., of Concord.

An estimated $2,000 damage was also caused to the sewage pumping station scheduled to be replaced next year by the new facility. All the controls were smashed, so that the flow of sewage to the American Canyon treatment ponds cannot be measured, reported plant supt. Sam Mutchek.

Without guages to guide them, operators will have to chemically treat the sewage by guessing the flow, said Mutchek. It will be several weeks before all the equipment is restored to working order, he estimated.

Investigating Napa County sheriff's deputies believe entrance to the site was gained during the night by persons who cut the lock on a chainlink gate. There was no night watchman on duty.

"They just stole from us all our power equipment," said Jenkins, who noted that a similar break-in Memorial Day weekend resulted in a $500 equipment loss. "This time they came in and took everything."

Because of the extensive vandalism throughout the site, Jenkins said the thieves were probably kids. Stealing keys to a pickup, the vandals crashed the vehicle into a pile of cement block and then ran it through surrounding fields until a tire fell off.

When completed, the pumping station will carry sewage from American Canyon to the new $11 million treatment plant near the Napa County Airport which will serve both Napa and American Canyon.

Index

Napa County Earthquake Drill Planned

If an earthquake similar to the 1906 temblor struck the Bay Area tomorrow, how would local emergency agencies respond?

That's the question the state Office of Emergency Services hopes to find an answer to June 25 in a drill involving emergency services in Napa, Yountville, St. Helena and Calistoga.

The operation will be entirely "in-house," according to Ronald Gunderson, county purchasing agent and local coordinator of the drill. "You won't even know it's happening."

Dubbed "BASE — '76," objectives of the exercise are to:

— Test response of county and city emergency staffs.

— Test emergency operations plans and procedures.

— Test adequacy of back-up communications systems.

The exercise scenario simulates a high-magnitude quake with widespread damage and many casualties. The nine Bay Area counties will be "affected," with the San Mateo area "hardest hit."

Gunderson says the scenario predicts little damage in Calistoga and St. Helena, just "shakes and broken windows."

Yountville and the west side of Napa would be hardest hit, with possible collapsed buildings, fires and traffic jams.

"BASE" is part of a continuing effort to ensure prompt response to an earthquake in the San Francisco Bay Area," said William Ward, regional manager for the California Office of Emergency Services.

It began in 1972, with the National Oceanic and Atmospheric Administration's "Study of Earthquake Losses in the San Francisco Bay Area."

This was followed by planning efforts — coordinated by the state, federal government, and Golden Gate University — involving the nine Bay Area counties, 92 cities and 500 special interest districts, as well as the federal government.

"Problems to be faced include utility disruption, fires, sanitation, medical, transportation, rescue, building inspection and relocation of survivors," Ward said.

Expected to play a major role is the state's Mutual Aid System, in which one jurisdiction calls for help from others, including the state and federal governments.

The scenario postulates about 10,000 dead and 40,000 injured in the Bay Area, according to Cecil Byrd, senior planner for the Office of Emergency Services.

How hard would Napa County be hit? "We can't forecast this in our scenario because we'd give our secret away," Byrd demurred.

(Continued on Page 2)

explained Zelma Long. It would take time before anyone truly understood the full extent of the impact that one simple little blind tasting, and its seemingly impossible results, could have on the world.

IN A BLIND RAGE that lasted a decade, King Menelaus of Sparta ultimately managed to conquer the walled city of Troy and retrieve his wife, Helen, and in so doing laid such waste to the city and its inhabitants that historians would for centuries question whether or not the famous walled metropolis had ever even existed.

So, too, the judges on that fateful May day in 1976 would have very much liked to lay waste to their ballots, to destroy every trace and memory of what would be seen by some as treachery, by others as heroism. Could they have retrieved their ballots, perhaps future historians would have been left to debate whether the Paris tasting, like Troy, was merely an endearing myth. Unlike the razing of Troy, however, the evidence of the tasting was well preserved, and while many grumbled about the results, no one could honestly and with integrity deny them. The American wines had won. The walls had come down.

Chapter 4

THE ODYSSEY OF THE VINES

There is no greater fame for a man than that which he wins
with his footwork or the skill of his hands.
—The Odyssey

Odysseus was one of the heroes of the Trojan War, a Greek king with blood on his hands and accolades to his name whom the bards and poets would sing and celebrate forevermore. Yet in his glory, he angered the gods, who in turn ensured that his journey home was arduous and harrowing.

None of the wines that had been sent to Paris would make the journey back, but those who hailed from the port from whence they'd sailed, *wine country*, had an Odyssey in front of them, a drawn-out and meandering peregrination to rival even that of King Odysseus.

It took Odysseus a decade to get home to faithful Queen Penelope, but it took even longer for the Napa Valley to find its way to becoming the world's premier wine region, and the path was similarly imperiled.

"WHAT DID I DO wrong?" asked an astonished Mike Grgich into the receiver of his telephone.

"Nothing, nothing at all," Frank Prial reassured him. "I just want to interview you for an article." There was a pause. "Actually," admitted Prial, "The article is *about you*."

More silence. Prial had informed Grgich that he was from the *New York Times*. Grgich knew what this was about, of course, but it still didn't make much sense. Winemakers didn't get interviewed outside of the *Register* or the *Star*—not unless their name was Tchelistcheff, at any rate. Furthermore, Grgich's former life made him somewhat wary of reporters. "Somebody from the press, back in Communist Yugoslavia where he grew up, if they called you, you knew you were in trouble," his daughter, Violet, later recounted.

"Mr. Grgich?" came the persistent voice on the other end of the phone.

"Yes, yes of course," replied Grgich, bemused. "Tell me when."

THE NAPA VALLEY WAS abuzz with life, even more so than before. Stag's Leap Wine Cellars and Chateau Montelena couldn't keep their wine from flying out the door. "The day after that story came out... people come running into the store with a copy of *TIME* magazine saying 'Do you have these wines? Do you have these wines?'" George Taber recounted. "Most of them didn't have the wines, but then they turned around and started buying California wines. Overnight it changed the reputation of California wines."

"It was a partnership then, the winery, and I wanted to get back as many bottles as I could for future tasting, and my partner said 'Warren, we should be *selling* the wine, not *buying* the wine,'" remembered Warren Winiarski, chuckling. Next door, at Clos du Val, Bernard Portet was having a similar experience. "All of a sudden, we had requests for wine from England, from Belgium—not from France," he said with a smile. "I knew my wine was beautiful. Only the French could teach me how beautiful it was." "The brilliant future of Napa Valley and American wine became obvious overnight," Bo Barrett would say later.

BACK IN FRANCE, ALL was not well. The unwitting tasters-turned-judges of this sip heard round the world were taking a sound beating from their incensed and unsympathetic colleagues. If they had selected American wines over French, surely it was a fault in their palates and not a reflection of the quality of the wines. It was unpatriotic, unjust, un-*French*! Pierre Brejoux, inspector general of the Appellation d'Origine Contrôlée Board, was under fire. The calls for him to resign his position were emphatic, repetitive and nearly successful. This was France; heads needed to roll.

Claude Dubois-Millot, sales director of *GaultMillau*, found his inexperience oft referenced after the event and his taste disparaged both publicly and privately. Aubert de Villaine had his partner at the Domaine de la Romanée-Conti, Lalou Bize-Leroy, accuse him of treachery and tell him flatly that he was "spitting in the soup." Pierre Tari, the mayor of Margaux, was asked to resign. Christian Vannequé was chastised by his boss and forbidden to blind taste wine again. Pierre Brejoux was nearly fired. Jean-Claude Vrinat of Taillevant became a pariah among his fellow restaurateurs. And yet, most of the wine-professionals-turned-judges, as unsuspecting of the results as they were undeserving of the abuse they suffered as a result, did their suffering in silence. Recalled Spurrier: "No one told me until thirty years on, when the ones who were still alive said: 'We can tell you now. It was ghastly.'" One of the judges, however, was not about to suffer, silently or otherwise.

A leading authority on wine and an experienced journalist and editor, Odette Kahn was well known as editor of the publications *La Revue du Vin de France* and *Cuisine et Vins de France*. Kahn refused to suffer the slings and arrows of outrage in the same manner as so many of her counterparts. Kahn had been infuriated and embarrassed at the results of the tasting and dramatically demanded her ballot back. Spurrier flatly told her, "No." In retaliation, Kahn spent a great deal of time after the tasting moving both her mouth and pen, insulting Spurrier and baselessly asserting fraud. This may have preserved her reputation and seems to have shielded her from some of the vicious attacks that her more honest contemporaries endured. Reportedly, Kahn refused to speak to Spurrier for an indefinite period after the event. She died in 1982.

Spurrier, the Englishman who had once endeared himself to French producers by learning their language and loving their wine, found himself suddenly persona non grata. "I was physically thrown out of the Romanet Cellars," Spurrier would later recall. "But since both the shop and the Académie continued to do well, I paid no attention," Spurrier asserted in his memoirs.

It was Steven's wife, Bella, who took the photographs of the event. "I was working for a photographer at the time," explained Bella. "I took the photos as practice, actually…then I got out of the way." Though she left before the tasting concluded, Bella's recollection of the event is one of the few firsthand accounts available. "They were all just sitting at the table, just as you see them in the photographs," said Bella. "I remember everybody sipping and slurping and writing notes," she laughed. "But I wasn't there for the great unveil. I only heard about that afterwards."

In the press, the French put on a good show. *Le Figaro* published a dismissive article, calling the results "laughable," while *Le Monde* published similar sour grapes, claiming that it "cannot be taken seriously." Of course, no journalists from either publication were present—only George Taber had bothered to attend. "If George hadn't been there, there'd have been no articles, and if my wife hadn't been there, there'd have been no photographs," Spurrier later asserted.

The subsequent French articles disparaging the results, thus, were not journalism but propaganda, a feeble attempt to un-ring a bell that originated in Paris and was resounding from sea to shining sea. The impenetrable walls that surrounded the myth of French enological superiority were no more.

BUT THE NAPANS FACED an uphill battle. While French wine law was deeply entrenched in history and culture, offering significant protections to the land, America's West remained wild in this way. While the Second World War was still being fought in Europe and the Pacific, those with a vested interest in the wine industry had formed their own band of brothers to help promote their trade and protect their interests.

The Napa Valley Vintners was originally formed by seven men and women invested in the commercial success of winegrowing in the valley. They included Fernande de Latour of Beaulieu Vineyard; Elmer Salmina from Larkmead; Robert Mondavi, representing his family and—at that time— the Charles Krug Winery; John Daniel Jr., who owned Inglenook; Louis M. Martini; Charles Forni from the Napa Valley Co-op; and Louis Stralla, who had been the one to tell Robert Mondavi that Charles Krug was up for sale. The group grew steadily from one decade to the next, successfully advancing the interests of the ever-expanding viticultural industry in the Napa Valley. By 2022, the web page of the Napa Valley Vintners boasted a membership of 539 wineries—if ever there were that many to begin with.

Early on, however, this small gang fought to find its way. The banks of the Napa River where the Wappo children once played games and grew into adulthood had become eroded and contaminated, and the Napans shared a general fear that their beautiful, riparian land might end up looking like the Silicon Valley. In 1966, they began an effort to protect the land by pushing for legislation that would prevent it from being divided into tiny lots for housing. The best and highest use of the land, they argued, was agriculture.

Warren and Barbara Winiarski, Jack and Jamie Davies and countless others went to war with development. A series of public hearings showed residents to be divided on the issue, but the vintners persisted. A developer who frequented the meetings in opposition to the proposed Ag Preserve, legislation that would restrict the use of large parts of the valley to agricultural purposes, showed up one day suddenly in favor. The reason? Gene Trefethen told him he'd buy his land, all of it, and at asking price—but only if the Ag Preserve passed. People rallied to protect the tiny valley, and in the end, they won. In 1968, the Ag Preserve was created, making the minimum parcel size in the valley twenty acres. Andrew Pelissa, the only grape farmer on the planning commission, was known to be politically conservative and remained tight-lipped on the issue, but he placed high value on the agricultural nature of Napa and voted in favor of the Ag Preserve. "All of a sudden, he was a pinko-liberal-dope-smokin'-commie for voting for it," recalled his son-in-law Ren Harris, "but people got over it." A few years later, the minimum parcel size would double from twenty acres to forty. The Napa Valley would not become the Silicon Valley—the vintners had made sure of it.

WHAT MIKE GRGICH AND Jim Barrett shared in mutual respect, they lacked in warm feelings, and while Grgich was once elated to own a small share of the now-world-renowned Chateau Montelena, his experiences in Yugoslavia made him crave the American dream even more. The year after Chateau Montelena's 1973 Chardonnay won the Judgment of Paris, vaulting the winery's reputation into the stratosphere, Mike Grgich left to strike out on his own.

"The Judgment of Paris had opened doors for me," recalled Grgich, "and through them came many people offering me work at their wineries. But for me, this would have been more of the same, of working for someone else." Grgich wanted to open his own winery, and while it was a difficult proposition, what he lacked in resources, he made up for with foresight and ambition. "I wanted to be on my own and make my wines the way my heart and soul told me to do it. For the first time since I had arrived in the Napa Valley, I finally felt close to attaining my dream," he recalled. Grgich sold his shares of Chateau Montelena for $50,000 and began looking for land to call his own.

Finding land in the Napa Valley wasn't terribly difficult. The Swiss chocolate company Nestlé had purchased the historic Beringer winery back in 1970 but soon discovered that the turnaround time and subsequent profit margins on wine were not on par with the bulk cocoa biz. Around the time Grgich was looking for land, Nestlé was selling off assets to keep its ledgers in the black. Grgich wanted two acres, but the Ag Preserve ordinance wouldn't allow it. Somehow, the ever resilient and resourceful immigrant cobbled together $100,000 and with it purchased twenty low-lying acres of wild grassland from Nestlé in Rutherford.

Meanwhile, Austin Hills of Hills Bros Coffee owned 150 acres of vineyards in Napa and paid Lee Stewart to make his wine at Souverain Cellars. Hills was looking to advance in the industry, and Grgich seemed the perfect partner. Grgich brought reputation and skill, while Hills provided a level of financial backing that Grgich needed. Austin and his sister, Mary Lee, joined forces with Mike, and Grgich Hills Winery was born. Despite the latter partners' suggestive surname, the land on the valley floor was consistent and level, and Grgich's young daughter, Violet, was less than thrilled by the transition. "I remember being very disappointed. We were on flat ground that was bare. There was no castle and no lake," she would later recall, only half-jokingly.

Grgich put the same amount of care into assembling his winery as he did into making his wines. At the four corners of what would become the footprint of the new winery, Grgich placed four bottles of wine: a 1958 Souverain Cabernet Sauvignon, a 1968 Beaulieu Georges de Latour, a 1969 Mondavi Private Reserve and a 1973 Chateau Montelena Chardonnay, four of the finest wines he had ever made, each one representative of a different stage in his journey on the way to achieving his American dream.

Mike Grgich and Austin Hills. *Courtesy Grgich Hills.*

The first year was a scary one for Grgich, who had left the relative comfort of being employed at Chateau Montelena to gamble everything on his own talents. He and Hills broke ground on the new winery on July 4, 1977, and found themselves racing the clock to have the winery constructed in time for crush. Grgich sought counsel from his old friend, Robert Mondavi.

"What's the problem, Mike?" asked Mondavi.

"I am building a little winery, and I have bought sixty tons of Chardonnay but I don't think I will be ready to crush it," Grgich responded.

"When did you break ground?"

"July four."

"Well, I broke ground on July [16], 1966 and I made it," Mondavi responded encouragingly.

Grgich remained unconvinced. Mondavi assured him he'd be fine but also jotted a handwritten note on a sheet of yellow legal paper assuring Grgich that, should his winery not be finished in time, Mondavi would crush and process his first vintage.

By September 5, the grapes were ready and so were the destemmer and the tanks. Grgich Hills lacked a roof, so the partners slung plastic tarps throughout the exposed rafters and celebrated their first crush in the unfinished building. Father George Aziz of the Catholic church in St. Helena blessed the grapes and the people alike, sprinkling them all in holy water.

A young Hispanic man approached Grgich one day while he was working. "Do you remember me?" the man asked. Grgich looked into his eyes and saw a little boy who had played at his father's feet at Beaulieu Vineyard decades before. That same boy, Grgich clearly remembered, had joined the team at Chateau Montelena in 1976 after completing his studies at UC Davis. Gustavo Brambila was one of the first Mexican Americans to graduate with a degree in viticulture and enology, and Grgich recognized many parallels between his journey and that of the man standing before him. "Of course I do, *Gustavo*," Grgich grinned. The pair would work side by side at Grgich Hills Estate for the next twenty-three years.

While Brambila worked for Grgich, the latter taught the former everything he could. "We would challenge one another," recalled Brambila, who invariably refers to his former employer as "Mr. Grgich" even all these years later.

"What's the acid on that wine?" Grgich would ask Brambila.

"Let me check," replied Brambila.

"That's not what I asked you," winked Grgich.

Mike Grgich, excited about the future of Grgich Hills. *Courtesy Grgich Hills.*

And so it went. Like Lee Stewart, Brother Timothy, André Tchelistcheff, Robert Mondavi, Zelma Long, Jim Barrett and everyone else who knew him, Gustavo Brambila marveled at the intuitive manner in which Grgich approached his craft. "Mr. Grgich had a self-awareness of wine, and all aspects of winemaking," recalled Brambila. "He didn't need to resort to analytics."

HIGH ON SPRING MOUNTAIN on the northwestern side of the valley in the Mayacamas range, Stu Smith was tending to the vines that sprouted at his bidding from the well-drained volcanic soils in his vineyards. Inside the winery that they had built themselves on the land first acquired by George Cook, Charlie Smith made the wine. They were a two-man operation with a small amount of experience, a few hundred acres, a natural spring and a breathtaking view of the valley floor below. As he walked the vineyards, Stu pondered the results of the recent Paris tasting. They didn't surprise him— he didn't build a winery in the Napa Valley because he doubted the ability of the land to yield extraordinary results. A thoughtful man who didn't mind the solitude of Spring Mountain, Stu woke up every morning and consumed half a dozen newspapers along with his breakfast before making the trek up the mountain from St. Helena. Stu had read that, with Spurrier's help, the

Vintner's Club in San Francisco had hosted a rematch of the Judgment of Paris the following year and that the results had again favored California. The more he read about California wines besting their European counterparts, the more convinced Stu became that it wasn't that odd at all.

By 1979, the Napa Valley was brimming with people who relocated to wine country to seek their fortunes in wine. Some sought new careers, some investments, some vanity projects. But those like Stu and Charlie—the old guard who, alongside Tchelistcheff, Stewart, Grgich, Barrett, Davies, Heitz, Winiarski, the Mondavis and many others, had beat the rush—simply continued to do what they'd been doing, making great wine in the beautiful place they felt fortunate to call home.

One Sunday evening during harvest, Stu remembers being up a ladder at the top of a tank when the phone rang. He climbed back down the ladder to get it; a friend spoke excitedly on the other end of the line. "My mom just got back from Europe!" he told Stu, "and she brought the *International Herald Tribune* with her!" Stu waited patiently for the story to get more interesting. After a moment, the friend blurted out, "You won! The Riesling competition! Stu, you and Charlie won!" In the future, the gravity of this information would not be lost on the Smiths, but in the moment, after yet another seventeen-hour workday, Stu just wanted to climb back up the ladder so he could finish his work and get home. "Great," he said wearily. "Send it to me?"

That tasting, put on in 1979 by *GaultMillau*—comparable to the Michelin Guide—and touted as the Olympiad du Vin, the Wine Olympics, was a truly international affair that far exceeded anything Spurrier or Gallagher ever envisioned. *GaultMillau* had declined to cover the Paris tasting but had certainly taken note of the results. The Judgment of Paris had raised questions. *GaultMillau* now sought to answer them.

In the Wine Olympics, hundreds of wines from around the world were tasted by expert judges, and the results again were stunning to those who had managed to convince themselves that the Judgment of Paris was somehow a fluke. If Napa had yet failed to capture anyone's attention, the Olympiad du Vin surely succeeded in doing so. As Stu's friend had attempted to impress on him, Smith-Madrone came in first in the world in Riesling, up against the best of Germany, France and others.

It wasn't only the Smiths who triumphed. "You must be mistaken," replied Janet Trefethen into the receiver, and then her mood darkened. If this wasn't a mistake, perhaps it was some kind of rude joke? "I assure you, I am not mistaken," said the voice on the other line. "Your 1976. It has been named the best Chardonnay in the world by *GaultMillau*." Janet was teeming with questions: When did this occur? Who organized it? And most of all: How in the hell did a bottle of her wine get to France? Nobody had contacted the Trefethens about entering a competition. The whole thing seemed strange, but if Chateau Montelena and Stag's Leap Wine Cellars could do it, then why not Trefethen? She thanked the caller and hung up the phone.

While the Smiths won the Riesling category and the Trefethens the Chardonnay, the top Sauvignon Blanc went not to Sancerre but to Sterling, another Napa Valley producer. Sterling looked Mediterranean from afar and was designed with input from Chalone's Dick Graff. A tram hauled visitors from the valley floor up to the winery. The year after the Judgment of Paris, Sterling sold to the Coca-Cola Company as part of the frantic scramble of outsiders staking a claim in the newly renowned Napa Valley.

After the results of the Olympiad du Vin were made public, Joe Heitz spied Stu Smith across the room at a meeting of the Napa Valley Vintners and approached him, extending his weatherworn hand. "Good going, kid," Joe said simply. "I remember that very fondly," Smith later recalled. "It was amazing to get that kind of respect from Joe Heitz, because I respected *him* so much."

ONCE AGAIN, BACK IN France, all was not well. Robert Drouhin, proprietor of Maison Joseph Drouhin in Beaune, felt personally affronted by the results of the Olympiad du Vin. Spurrier's tasting could be written off, he reasoned, irritating though it was, but for *GaultMillau* to follow suit? This nonsense had to stop immediately, Drouhin thought, and he himself would be the one to put an end to it. Drouhin wrote a letter to *GaultMillau* in which he threw down the proverbial gauntlet. "As a French person, Burgundian, winegrower, cultivator of great wines, I was passionately interested in the 'Wine Olympics' that you organized, and terribly disappointed by the results.…I am ready to challenge the winners of your Olympics in the Pinot Noir and Chardonnay categories with a selection of Burgundy wines from Maison Joseph Drouhin."

On January 8, 1980, only a few months after the Olympiad du Vin, Sisyphus got back behind his boulder and continued to shove it up Mount Wine. Much like the previous tastings in 1976 and 1979, Drouhin assembled a panel of reputable judges to evaluate the wines. As he expected, Drouhin's Burgundian Pinot Noirs outperformed those of Switzerland and Macedonia, though in a very close second place—and ahead of five out of six of Drouhin's—was a curious entry from a virtually unknown wine region. The second-place wine was a 1975 Eyrie Vineyards South Block Pinot Noir from Dundee, Oregon. Crafted by David Lett, another graduate of UC Davis, this extraordinary Oregon offering captured Drouhin's attention and imagination.

In the white wine category, Drouhin got a nasty shock when the 1976 Trefethen again came in first and the 1975 Freemark Abbey came in third, leaving Drouhin's exceptional white Burgundies behind. To his credit, the crestfallen vintner later called the Trefethen 1976 "the yardstick by which all other Chardonnays must be measured," though many critics felt the 1977 Trefethen was better still. Robert Drouhin eventually went on to purchase land immediately adjacent to the Eyrie Vineyard in Oregon, establishing Domaine Drouhin.

Warren Winiarski would later joke about the Judgment of Paris: "There were a number of French astronomers at that tasting, and they discovered that the sun does not only go around France." As if to prove his point still further, the wine exploits of 1980 neither began nor ended in Germany's neighbor to the west. Foreshadowing what was to come, at the Orange County Fair, a panel of judges that included Winiarski awarded a gold medal to a Chardonnay from a brand-new winery in Rutherford called Grgich Hills. "It was the first gold medal for Grgich Hills," Grgich later remarked. There were more to come.

Warren Winiarski and Steve Spurrier. *Courtesy Warren Winiarski.*

Soon after, the city of Chicago staged the Great Chicago Chardonnay Showdown. A wine writer for the *Chicago Tribune* named Craig Goldwyn worked with wine merchants across the Chicagoland metro to put on a tasting of 221 Chardonnays from around the world. The panel of twenty-five judges diligently worked their way through rivers of golden grape juice until at last they arrived with their winners. Twelve of the top twenty were the recognizable French names in Chardonnay, from Meursault Premier Cru to Petit Chablis to Macon Villages to Batard-Montrachet, Puligny-Montrachet, Chevalier-Montrachet, and so on. Of note, however, were again some exciting American success stories. The 1978 Dry Creek Vineyard came in eighth, a win for Napa's neighbors in Sonoma. Inglenook, then the property of Heublein, was eleventh. It was the top five, however, that again gave the entire world of wine something to talk about.

In fifth place in the Great Chicago Chardonnay Showdown was the 1978 Chateau Montelena. Montelena had become a household name, and the success of its Chardonnay at such an event could hardly catch anyone off guard. Jerry Luper was then the winemaker, having been hired away from Freemark Abbey. Luper was also mentoring a young Bo Barrett in the cellars.

In fourth place was another Sonoma producer, the historic Simi Winery, with its 1977 made by MaryAnn Graf. In third, a 1974 from Joe Heitz, who continued to make the sort of wine his wife loved. In second place, the French snuck in a wine, the amazing 1978 Beaune Clos des Mouches from the newly converted believer in American wines, Joseph Drouhin. In first place was the 1977 Grgich Hills Cellar Chardonnay. Mike Grgich had done it again.

That the 1977 was the first vintage of Chardonnay from Grgich's new winery did not escape enthusiasts. "The first wine his new winery has produced has again topped the French competitors," crowed Craig Goldwyn in the *Chicago Tribune*. "Grgich did it again," wrote Jerry Mead in the *Vallejo Times-Herald*. Shortly thereafter, Grgich was given the nickname King of Chardonnay. For a Croatian immigrant who had desperately fled an oppressive regime and eventually arrived in Napa with no money to accept a seasonal position at a small winery, this moniker felt like almost too much. "Imagine that—to be called a king! I felt on top of the world," reflected Grgich.

LIFE REMAINED A WHIRLWIND for the seemingly stir-crazy Spurriers, who spent increasing amounts of time in California. In the fall of 1983, Steven Spurrier acquiesced to serving as a wine judge at the Sonoma County Fair, and while he was in the area, he took the opportunity to visit Napa again. While there, he dropped by Grgich Hills unannounced. "I pushed open the large oak door to the barrel cellar and there was Mike, one of the great heroes of California wine history, telling a group about the Paris Tasting," Spurrier later recalled. "As he mentioned my name, he saw me at the door and was so surprised that he even took off his signature blue beret."

A few years later, Spurrier agreed to help hold a tenth anniversary tasting of the same wines and vintages that were tasted in 1976. Some had argued that the California wines only performed as well as they did because the French wines took longer to mature and that the superiority of the French wines would best be demonstrated by their longevity. Others argued that the twenty-point system Spurrier and Gallagher used in the original tasting, a system of scoring that was prevalent at the time, was an unfair method, as well.

So, in 1986, with the same wines and the same vintages but a new and improved scoring method and a new panel of judges (the old ones were not about to subject themselves to this a second time) that included Robert Finigan and other prominent names in wine, the Paris tasting was reenacted in New York. Only the red wines were judged; while white wines, in particular Chardonnay, can age well for many years, they are generally not thought to improve over a period of decades in the way that red wines are. The results of the tenth anniversary tasting are as follows:

First: Clos du Val 1972 (California)
Second: Ridge Monte Bello 1971 (California)
Third: Château Montrose 1970 (France)
Fourth: Château Leoville Las Cases 1971 (France)
Fifth: Château Mouton Rothschild 1970 (France)
Sixth: Stag's Leap Wine Cellars 1973 (California)
Seventh: Heitz Martha's Vineyard 1970 (California)
Eighth: Mayacamas 1971 (California)
Ninth: Château Haut-Brion 1970 (France)

California reigned victorious again. Of note, while three Bordeaux houses provided their wines to Spurrier for the tasting, Château Haut-Brion refused to participate, so Spurrier purchased a bottle of their wine in New York, which he remarked later had likely been improperly stored. Freemark Abbey declined to be involved, feeling that by then their 1969—the oldest vintage to have been tasted in Paris—had faded. Recalling Clos du Val's victory, Bernard Portet was understated. "That was good," he smiled with a raise of his eyebrows and a slight shrug.

Steven Spurrier's reputation in America was sterling, but the Spurriers were soon short on silver. A combination of factors contributed to what ultimately became their financial hardships. "He wasn't a very good businessman," Bella Spurrier conceded much later in life. "He had wonderful ideas, but they were mostly for other people."

The wine shop wasn't making much money, if any. Spurrier was popular among expats, while most of the French were either indifferent or nursing

1971
VINTAGE

Cabernet Sauvignon
WINE

GRAPES:
Mayacamas – 3 tons
Draper – 20 tons
Rhodes – 1 ton Merlot (Zinfandel Ln. – Zinfandel Assoc.)
VINEYARD

SUGAR 23.7 *ACID* 0.70 gm/100 ml *pH*

Alc. – Under 13%

WINE:

WOOD AGING \2 Yr. American 500 – 1000 gal. 1 yr. French oak

BOTTLE AGING Summer 74 Released Sept. 75

FINING No

FILTERING Light

BLEND 3% Merlot – Adds Complexity

YIELD 1274 5th 103 10th 128 Mag.

COMMENTS Bob's: Somewhat rough and very closed in. True characte

will not be evident before 1980. Personal preferences in balance of

fruit, tannin, acid, etc. can vary the criteria for such a judgement.

LONGEVITY Mid to late 1980's. May self-destruct before 1995.

Malolactic: Yes

Comparison: Quite unique but a little like 68 and 73

/87 *still getting better; best earl–*

Winemaker's notes on the Mayacamas Cabernet Sauvignon 1971, dated 1987. "Still getting better." "May self-destruct before 1995." *Photo by author.*

a grudge. Steven Spurrier was also all over the place, and ironically, while he had one of the most sophisticated palates in the industry, his attention to detail in other realms was lacking. "He hadn't really taken into account the expenses of running a wine shop in Paris," Bella remarked. "He ran it very well in terms of wine and selling wine, but he wasn't interested in the nitty-gritty like electricity and rent."

In addition to the wine shop, Spurrier had become a Parisian restaurateur. When he noticed that one of his leads was living far beyond his means, Spurrier brought in Christian Vannequé, who had served as a judge at the Paris tasting, to "up its game" as well as to keep an eye on the employee he mistrusted. Over time, and despite his efforts to correct the situation, the finances failed to improve. When Spurrier confronted Vannequé, the latter astounded him by admitting that "a little had been taken off the top," quickly defending the statement by arguing that this practice was common in the restaurant business. When Spurrier in turn asked, "Where is your sense of honor?" Vannequé replied flippantly, "I'm sitting on it." "Very French," Spurrier later reflected.

The more Spurrier looked into things, the worse they turned out to be. Even his bookkeeper was defrauding him. The bottom fell out when François Mitterrand was named president of France. "There was a huge revolution in France, and in came President Mitterrand, who increased interest rates dramatically," remembered Bella Spurrier. In a failed attempt to combat unemployment, Mitterrand oversaw a meteoric rise in interest rates that peaked around 18 percent. "We had money borrowed, so that pretty much wiped us out," said Bella.

What wasn't wiped out by the failed policy of astronomical inflation rates in France managed to disappear in other ways. The French authorities falsely accused the Spurriers of tax evasion, claiming that Steven's claimed losses of three million francs were actually profit he had smuggled out of the country. While this was baseless, it was left to the Spurriers to prove it, and the means of doing so were expensive. Shortly after, Spurrier attempted to remove his assets from another business in which he was invested, only to have his "partners" refuse to pay him on a technicality—the documents that guaranteed Spurrier four million francs had never been notarized. To make matters still worse, despite being swindled out of his compensation from the business, he was still on the hook for the same business's debts. "This was a devastating blow," Steven later wrote.

In 1990, when Steven was forty-nine, he and Bella Spurrier returned to London. The debonair, charming, happy-go-lucky man who had done so much for the world of wine—and, in particular, American wine—found himself in a place he could never have imagined during his carefree upbringing at Marston Hall in Derbyshire. Recalled Spurrier later, "I was not only broke, but also deeply in debt, and the future looked unimaginably bleak."

As THE POPULARITY OF the Napa Valley soared following the Judgment of Paris, more and more people wanted a piece of the action. Even before the Judgment sent shock waves through the world of wine, Robert Mondavi and the Baron Philippe de Rothschild, the proprietor of Bordeaux's esteemed Château Mouton Rothschild, had been scheming about a new world–old world collaboration. This came to fruition in the late 1970s in a project called Napa Medoc, later Opus and finally renamed Opus One. "We called it 'Bob's folly,'" recalled one longtime employee. Ironically, that's what many had called Mondavi's original winery just across the highway, as well.

The lawsuit between the Mondavi brothers concluded in 1978, though the feud would persist long after. Robert was granted some 250 acres in the lawsuit and used the money to buy hundreds more. Around that time, Gunther Detert, who had a stake in the To Kalon Vineyard, was attempting to convince Robert to resurrect the historic vineyard's name. In response, Bob asked "Why would I put an odd Greek word on the label when people don't know who Robert Mondavi is yet?" Name recognition wouldn't be a problem for Mondavi much longer. The first vintage of Opus One, 1979, was released alongside the 1980 in 1984. By then, Robert Mondavi was fast becoming the face—and name—most closely associated with American wine.

"Take this home and tell me what would make it better," said Warren Winiarski, holding out a bottle of wine. Michael Silacci knew Winiarski by his reputation, and he wanted to work for him. For his part, Winiarski was going to make the young man earn it. Sometimes, it was homework, like the bottle of wine he was presented with. Other times, it was a long sit-down or a walk through the vineyard. In all, Michael Silacci estimates he interviewed with Warren Winiarski ten times before being offered the position of assistant winemaker. Silacci turned it down. He wanted to be the *associate* winemaker and work with the estate wines. In part on the recommendation of Dorothy Tchelistcheff, Winiarski hired Silacci as his associate winemaker. After working for Winiarski at Stag's Leap Wine Cellars and later André Tchelistcheff, who had agreed to take Heublein's money as a consultant at Beaulieu, Silacci would eventually land at Opus One.

Left to right: Robert Mondavi, Baroness Philippine de Rothschild and Baron Philippe de Rothschild. *Photo courtesy Opus One.*

But it wasn't just the big names in wine like Mondavi and Rothschild who were penitently planting the valley to vines and erecting cathedrals to fermentation in the years following the Judgment of Paris. Dennis Groth left his position as an executive at Atari in search of a quiet life in wine. He and his wife, Judy, hired realtors Ren Harris and his partner, Jean Phillips, who showed them one property after another. After each property, Dennis would get out his pencil, start crunching the numbers and talk himself out of buying it. On New Year's Eve 1981, Ren and Jean showed Dennis yet another property, this one very close to Opus One in Oakville. When Dennis, predictably, took out his pencil, an exhausted and beleaguered Jean took it from him and broke it in half. "I'm sick and tired of your damn pencil," she told him. "It's New Year's Eve and I need a commission. Just buy it!" "I said OK," Dennis later recalled with a smile.

Having parted ways amicably with Joe Heitz, Nils Venge and his father-in-law opened Saddleback Cellars. To foot the bills of the endeavor, Nils took a day job working for Dennis Groth at his new winery less than a mile away. Together, they made history when their 1985 Cabernet Sauvignon became the first American wine to earn a one-hundred-point score. There could no longer be any doubt: American wines were among the world's best.

Elsewhere in the valley, up in Angwin on a mountain named for John Howell, who set up shop on George Yount's second land grant, Rancho La Jota, nearly a century before, Randy and Lori Dunn purchased fourteen acres, five of which were planted to Cabernet, and opened Dunn Vineyards. Farther north in Knight's Valley, having failed to land Peggy Lee, Sir Peter Michael purchased land and opened his eponymous winery. Mary Novak and her husband, Jack, purchased the Old Kraft Winery in St. Helena in 1972, and in 1982, they released their first wine, having renamed the estate Spottswoode. All around wine country, more and more wineries were springing up in the years following the Judgment of Paris.

Bottle of Groth Reserve Cabernet Sauvignon 1985, the first hundred-point wine made in America. *Photo by author.*

While many new wineries opened, others simply changed hands. Sometime in the early 80s, Coca-Cola decided to get back in its lane and sold Sterling Vineyards to Seagram, which didn't. Sunny St. Helena, in which Cesare Mondavi had been a partner even before Robert talked him into buying Charles Krug, was sold in 1983 to the Christian Brothers, who renamed it Merryvale. Lee Stewart sold Souverain Cellars to Pillsbury back in 1970, and the doughboy promptly destroyed its reputation for making fine wine. In 1986, Nestlé added the historic winery to Beringer in its portfolio. Shortly thereafter, Christian Brothers was sold to Heublein, joining Inglenook and Beaulieu. It became increasingly evident over time that most large companies had no clue how to make money in the wine business, but this didn't keep them from buying up historic wineries in the years following the Judgment of Paris.

It wasn't just the Napa Valley that was being bought up, however, and those wineries that were linked to the Paris tasting were in high demand. In 1987, Ridge Vineyards was sold to Otsuka Pharmaceutical, a Japanese company that went out of its way to retain Paul Draper and insisted that nothing change at the winery, save for who was writing the checks. Farther south and a few years later, not to be outdone by the Bordelaise neighbor who shared his surname and had partnered with Robert Mondavi, Baron Eric de Rothschild of Château Lafite Rothschild announced a partnership

Laborers tending the vines at Ridge Monte Bello. *Photo by author.*

with Chalone, which had become publicly traded in 1985, in which each winery acquired a 6 percent ownership share in the other.

Wine had captured the imagination of Americans, who suddenly, after years of slurping sweet jug juice, found fine wine interesting again. Beginning in 1981 at Spring Mountain Vineyard, a soap opera called *Falcon Crest* was filmed. The show was not only set in wine country, it was *about* it in some ways, too. While the show aired, Spring Mountain Vineyard produced and marketed a "Falcon Crest" wine. The 1980s marked a period of experimentation, gambles, scrambles and explosive growth in wine country. The attention of the world had been ensnared by the lusty allure of grape must and the promise held by fertile land and possibility. The last episode of *Falcon Crest* aired on May 18, 1990. The world's interest in the Napa Valley would persist forever after.

ODYSSEUS'S JOURNEY FROM TROY back to his native Ithaca was long and perilous. Similarly, the journey of the Napa Valley from a once-somnolent agrarian paradise to becoming one of the world's premier wine regions not only took time but also consumed the lives of so many who were a part of the exciting excursion into the future. The rich and diverse soils of Napa were fertilized by the footsteps of a new wave of pioneers who dedicated their lives to planting and protecting the area. At last, Odysseus would reach the shores of Ithaca, and at last, after generations of toil and triumph, the Napa Valley would reach its destination as well. By the beginning of the twenty-first century, there could be no dispute that the Napa Valley was one of the—if not *the*, singular—premier wine-growing regions in the world.

On his way home, Odysseus lost many friends and fellow soldiers as he struggled to reach his destiny. Likewise, along the way, the Napa Valley bid one fond farewell after another to the great pioneers who had transformed it on the journey.

"On Tuesday, April 5, 1994, André Viktorovich Tchelistcheff, who had cheated death on more than one occasion in the past, took his final breath," wrote a biographer. At the funeral, there was reportedly more laughter than tears, as the man who had transformed not only the Napa Valley but also the entire American wine scene was remembered. In eulogizing the innovative White Russian, Legh Knowles of Beaulieu dubbed the man Maestro, a moniker that would all but replace his difficult-to-pronounce surname in the years to come.

In the spring of 1998, Jack Davies passed away at his home at Schramsberg, in the beautiful old home where Jacob Schram and his wife had entertained Robert Louis Stevenson. Not long after, in St. Helena, Louis P. Martini, son of Louis M. Martini, died of cancer at the age of seventy-nine. A few years later, on a brisk Saturday in December, Joe Heitz, who had never fully recovered from a stroke, passed on. In the years to come, Robert Mondavi, Jim Barrett, Peter Mondavi and countless other innovators to whom so much is owed would depart the Eden of the earth for the one that awaited them after. Each left their own indelible mark, and their legacies endure in that small and beautiful space between the Mayacamas and the Vacas.

Chapter 5

THE FAIREST ONE

Like that star of the waning summer who beyond all stars rises bathed in the
ocean stream to glitter in brilliance.
—The Iliad

Chef Thomas Keller opens his dream restaurant, The French Cinema, in West Hollywood. 'I plan to source the freshest ingredients from throughout southern California, and I need a client base that is willing to spend the money for refined food,' he says. The restaurant closes three years later and Keller takes a job as executive chef at the Ritz-Carlton."

In an article for the web page Wine Searcher in 2015, wine writer and editor W. Blake Gray speculated about what the world of wine might look like had the Judgment of Paris never occurred or if George Taber hadn't been there to cover it for *TIME* magazine. Gray's depiction, though purely speculative, is delightfully dystopian and equally grim.

Thomas Keller opened the French Laundry (not "Cinema") in Yountville for the reasons described in Gray's fictional portrayal, and not only is it still open, the French Laundry also boasts three Michelin stars and has three sister restaurants also owned by Keller in the same little town George Yount established as Sebastopol more than 150 years ago. These include a Mexican restaurant and another famous for fried chicken. The culinary legacy of the French Revolution is alive and well in Yountville.

"In 1976, the New World did not exist in France," said Steven Spurrier matter-of-factly. But while the French were extremely reluctant to be open-minded toward wine made in other parts of the world, the reverse was certainly not the case.

"Forty years ago in Washington State, there were twelve wineries. Forty years ago in Oregon, there were fewer than twelve wineries. Today in Oregon, there are seven hundred wineries. In Washington, nine hundred wineries and sixty thousand acres of grapes," said an enthusiastic Ted Baseler, representing Chateau Ste. Michelle at a fortieth anniversary celebration of the Judgment of Paris. "The transformation from this event was clearly not just Napa, not just California, and in fact around the world people drew this inspiration from this event, it was so tremendous."

Robert Mondavi called the Paris tasting "a victory for Napa, for California, and for all of North America." For possibly the first time in his life, Mondavi was underselling. In reality, the effects of the Paris tasting extended far beyond the North American continent to impact the entire world.

For one thing, the world began to drink more wine. The share of wine in total alcohol consumption may have stagnated or even declined somewhat in places that traditionally consumed wine, but elsewhere, enthusiasm for fermented grape juice was on the rise. From 1976 to 2016, the percentage of wine consumed more than doubled in Australia, Canada, the Netherlands and Mexico and more than tripled in New Zealand and Sweden. The growth was even more extreme in other markets. The United Kingdom and Ireland consumed nearly five times as much wine in 2016 as they had four decades prior, while in China and Japan, the growth was nearly tenfold.

Beyond consumption, of course, many nations began making wine commercially, such as China and Rwanda, while many more began exporting it. Perhaps the single greatest impact of the Judgment of Paris is best viewed today in the grocery store. On shelves that once offered mostly French wine with some European and American wines mixed in, today one can almost certainly find Argentine Malbec, Chilean Cabernet Sauvignon, Australian Shiraz, New Zealand Sauvignon Blanc, South African Pinotage, Lebanese red blends, Israeli Chardonnay and more.

The Napa Valley changed the world, and the world returned the favor. The Napa River still traced its reposeful line between Mount Saint Helena and the San Pablo Bay, and the sun still rose on the Vacas and set on the Mayacamas day in and day out. Everything else, however, seemed altered.

While the best efforts of conservationists—professionals as well as private citizens who sought to protect the valley—had managed to preserve large swaths of land, most of it for agriculture and some as untouched preserve, the home of the Wappo would have been all but unrecognizable to its original inhabitants. What had once upon a time been a quiet agrarian setting had been developed, paved and populated. It was certainly still beautiful, though substantially less tranquil, and awash with the trappings of unimaginable wealth. Those who had farmed grapes before the Judgment were still there, but now they were surrounded by billionaires and retired executives who had noted in the wine industry a certain *je ne sais quoi* and eagerly forked over small fortunes for a plot in paradise to call their own.

Between 1970 and 1980, the pre- and post-Judgment eras, the population of Napa County grew more than 25 percent, from just over seventy-nine thousand to just over ninety-nine thousand in that ten-year span. During that same decade-long period of time, and not by coincidence, American wine consumption nearly doubled, and what's more, Americans also began drinking more domestic wines. The population growth of Napa County continued to swell intensely for the next few decades, until in 2010, the tiny space seemingly reached critical mass. With farmland protected, wildlife preserves in place and the cost of land and housing pricing all but the inheritors and the wealthiest out of options, the population was at last arrested near 136,000 inhabitants, and it remains roughly that to this day. Those who owned houses and could afford to purchase land at exorbitant prices were fine, of course, but many who worked the land, waited tables at the increasing number of Michelin star restaurants or staffed the ubiquitous tasting rooms that dotted the hillsides and the valley floor found themselves moving over the mountains to Fairfield, to Santa Rosa, to any place where they could find an apartment they could afford to rent. And the most important laborers, those whose skilled and experienced hands deftly wielded the sickle-shaped blades that harvested grapes every autumn—well, when the harvest was over, most of them would return to Mexico. To live in California, let alone the Napa Valley, on the wages of a migrant laborer? *¡Olvídate de eso!* (Forget about it.)

Population was not the only thing to spike alongside the cost of fermented grape juice in the years following the Judgment of Paris. Land, perhaps the most finite commodity in the world, was in increasingly high demand. In 1943, Cesare and Rosa Mondavi purchased the Charles Krug Winery and the 147 acres around it for $75,000. By 2017, the average price of a single acre of land in the Napa Valley exceeded $300,000, and even defunct wineries sold for millions—if a person could find one for sale in the first place, that is.

Meanwhile, grapes from the To Kalon Vineyard, now an amalgamation of land owned in part by Constellation Brands as part of its acquisition of the Robert Mondavi Winery and in part by Andy Beckstoffer and other private individuals, could easily fetch $25,000 per ton. The vineyard was by then planted almost entirely to Cabernet Sauvignon, of course.

The manner in which Constellation came to own To Kalon and the Robert Mondavi Winery is contentious. Seeking to raise liquid capital, Mondavi examined Chalone, which by then was publicly traded. Mondavi knew Richard Graff at Chalone quite well. Along with Julia Child and others, they had established the American Institute of Wine and Food together. Tragically, Graff died in a small airplane accident in January 1998.

Mondavi determined that the best path forward was to sell shares. He pushed hard on his children, who stood to inherit the winery, to acquiesce to the stock plan on the condition that the family maintain controlling interest. MOND hit the NASDAQ at $13.50 per share. Despite annual sales of $150 million, the stock soon plummeted to under $8. "Just imagine how I felt," Mondavi said glumly. But the worst was still to come. Soon, controlling interest was wrested away from the Mondavi family. According to one longtime Mondavi employee, "Bob's friends chose to go for the money and screw him." Overnight, Robert Mondavi's dream no longer belonged to Robert Mondavi. It belonged to Constellation.

A few years later and somewhat predictably, a third iteration of the Paris tasting was hosted, thirty years to the day from the original, again using only the red wines and the original vintages. Spurrier again was involved in putting on the tasting, and what's more, he included Christian Vannequé, Patricia Gastaud-Gallagher and Michel Dovaz, three of the original judges. The major difference this time was that there were two panels, one on either side of the Atlantic. The London tasting was held at six o'clock in the evening at Berry Bros & Rudd, while the Napa tasting was held at Copia at ten o'clock in the morning; their scores were ultimately added together. While Paul Draper had now assumed the position previously held by Bernard Portet and Warren Winiarski, the judges came to essentially the same conclusions as before:

First: Ridge Monte Bello 1971 (California)
Second: Stag's Leap Wine Cellars 1973 (California)
Third (tie): Heitz Martha's Vineyard 1970 (California)
Third (tie): Mayacamas 1971 (California)

Fifth: Clos du Val 1972 (California)
Sixth: Château Mouton Rothschild 1970 (France)
Seventh: Château Montrose 1970 (France)
Eighth: Château Haut-Brion 1970 (France)
Ninth: Château Leoville Las Cases 1971 (France)
Tenth: Freemark Abbey 1969 (California)

In his usual polite manner, Spurrier claimed to be surprised by the results, though it would be reasonable to doubt whether or not he was sincere, given that the results had never really varied. The BBC reported that the French producers might be "rather miffed" by the results, of which wine writer Jancis Robinson later reflected "'Miffed?' Pure Spurrier."

Following in the footsteps of Moët & Chandon—or, perhaps more accurately, those of Georges and Fernande de Latour—there came a French invasion. Champagne Taittinger opened Domaine Carneros. Champagne Mumm opened Mumm Napa. Christian Moueix of Bordeaux's Château La Fleur-Pétrus and Château Trotanoy had taken a recommendation from Robert Mondavi in the early 1980s and gone into partnership with Marcia Smith and Robin Lail—the daughters of John Daniel, who once owned Inglenook—on the historic Napanook Vineyards, where George Yount had planted the first grapevines in the valley. Mouiex opened Dominus Estate in Napa shortly thereafter.

Jean-Charles Boisset is a charming, flamboyant playboy with a roguish smile whose roots run deep in Bourgogne. Perhaps the only character in this story who could have outdressed Steven Spurrier, Boisset purchased Agoston Haraszthy's Buena Vista Winery in Sonoma, the old train depot in Calistoga, the grocery store in Oakville and more. Boisset married Gina Gallo, winemaker and heiress to the Gallo wine empire, cementing the couple's place as quite possibly the most formidable power couple in wine. Araujo Estate Wines, St. Supéry Estate Vineyards & Winery—one after the next, Napa wineries were gobbled up. "If you can't beat them, *buy* them" became the apparent mantra of the moneyed French wine aristocracy.

At the same time that the French and others were busy buying up Napa, those in Napa—always one step ahead—were investing elsewhere. Robert Mondavi was among many Napa producers with projects in South America. Mike Grgich was preparing to open a winery in his native Croatia. Jerry Luper, formerly of Freemark Abbey and later Chateau Montelena, had gone to Slovakia but soon decided he preferred Portugal. In each new place, Napa's winemakers found untapped potential, and they knew precisely what to do with it. "I am very happy here in the Douro making table wines and Ports," said Luper in an interview. "This is where Napa Valley was 20 years ago. It's like starting over again. I really lucked out." For her part, Zelma Long was asked to give a talk on winemaking in South Africa, and she fell in love with it and began making wine from Bordeaux varietals. The Napans knew that what they had was amazing, yet they were determined not to make the same mistakes as the French by overlooking the rest of the world.

About a quarter of the Napa Valley is owned today by a combination of foreign investors, including those from England, Switzerland, Australia and Japan in addition to France. Among them, Diageo is a UK-based company that owned Beaulieu as well as a handful of other wine holdings and, well, Burger King. Then Foster's, "Australian for beer," purchased Beaulieu, as well as Beringer, Stag's Leap, Sterling and others, rebranding itself Treasury Wine Estates, presumably to distance itself from cheap lager, while Warren Winiarski's Stag's Leap Wine Cellars is now owned jointly by Chateau Ste. Michelle of Washington State and Italian wine magnate Antinori. With so much outside interest, the Napa Valley fast became a seller's market, and many longtime residents who were wary of the impending changes took advantage of the opportunity to get while the getting was good.

Far from immune to the world around them, most of the wineries that were featured in the Judgment of Paris changed hands over time. Swiss billionaire Jacqui Safra purchased Spring Mountain Vineyards and numerous adjacent properties, including Chateau Chevalier, La Perla and the Draper home around the time that *Falcon Crest* went off the air. Safra got to work excavating the caves his predecessor had dynamited in a drunken attempt to impress the Beringer brothers but mostly governed from afar. Mayacamas became the property of the Schottenstein family in 2017. Among their many contributions, the Schottensteins added extraordinary non-*mevushal* kosher wines to the historic producer's portfolio. Freemark Abbey was purchased from an investment group by Sonoma's Jackson family in 2006. Chalone was rescued from Diageo by Robert Foley in 2016. Veedercrest appears to be the lone winery from the event that managed to go defunct on its own. Someone apparently still renews the web page annually, but it goes without updating and remains something of a mystery. "I believe [they] stopped producing," said one industry expert.

While the land and what was built on it clearly held tremendous value, the proverbial blue sky of the industry was selling for a fortune as well. In

1965, Davis Bynum dipped his daughter's foot in ink, put it on a wine label and called the wine Barefoot Bynum Burgundy. When someone offered to buy the label—just the label, with no winery or vineyards associated with it—Bynum thought that was nuts but gladly took their money. Since then, buying labels—labels alone—has become "a thing" in the industry—and a profitable thing at that. In 2010, winemaker Dave Phinney sold his Zinfandel label "The Prisoner" to Huneeus Vintners for a reported $40 million. Huneeus amped up production, doubling it from 85,000 cases a year to 170,000, then turned around and sold the brand to Constellation, the company that owns Robert Mondavi, in April 2016 for $285 million. Two months later, in June, Phinney sold Orin Swift Cellars to E&J Gallo for a reported $300 million. Back in 2002, when Gallo purchased the Louis Martini Winery, the acquisition came with hundreds of planted acres of vineyards, a tasting room, a massive production facility, a hundred-year-old reputation for quality and more. Orin Swift came with a small tasting room on St. Helena's main drag, but it was really just the name that Gallo was paying for. After that, Gallo went on to purchase countless other brands, and while hard figures are hard to come by, many guesstimate that Gallo produces more than 1 percent of the world's wine.

In the summer of 2007, Warren and Barbara Winiarski, having contributed so much to the Napa Valley and to the world of wine, sold Stag's Leap Wine Cellars—which boasted Winiarski's winery and world-class vineyards as well as an iconic label—for $185 million. Asked later about the decision, the philosophic winemaker, of course, quoted a poem, "A Lecture upon the Shadow" by John Donne:

> *Except our loves at this noon stay,*
> *We shall new shadows make the other way.*
> *As the first were made to blind*
> *Others, these which come behind*
> *Will work upon ourselves, and blind our eyes.*
> *If our loves faint, and westwardly decline,*
> *To me thou, falsely, thine,*
> *And I to thee mine actions shall disguise.*
> *The morning shadows wear away,*
> *But these grow longer all the day;*
> *But oh, love's day is short, if love decay.*
> *Love is a growing, or full constant light,*
> *And his first minute, after noon, is night.*

Warren and Barbara Winiarski enjoying a glass of wine. *Courtesy Warren Winiarski.*

To put it another way, Winiarski added simply: "Everything has its time."

But whether the wineries were being purchased by foreign companies or foreign nationals or, instead, by American corporations or simply by wealthy Americans, the common denominator was simple: there wasn't a square inch of earth or adhesive in the Napa Valley that somebody didn't want to acquire, and this drove prices even farther through the stratosphere. In the end, the hyper-gentrification of the bountiful yet undersized valley soon came to mean that the Adams and Eves whose entrepreneurial gambits and unshakable work ethic had reestablished this Eden after Prohibition could scarcely afford to live there any longer.

STILL, IN THE PUREST expression of idealism that Lady Liberty could muster, it wasn't impossible for the "little guys" to make their way in wine country. Gustavo Brambila, the son of immigrants, himself an immigrant, had risen to the top of the barrel. Born in San Clemente, Jalisco, Mexico, he arrived in Napa a stranger in a strange land that was often unfriendly to those perceived as outsiders. After a short stint at Cheateau Montelena followed by a long career with Mike Grgich, he set out to start his own label with Thrace Bromberger. By 2013, Bromberger had moved on, and the by-then-well-known Gustavo Brambila rebranded his winemaking enterprise Gustavo Wine.

Gustavo Brambila is grizzled and thickly mustached, resembling the Most Interesting Man in the World from the beer commercials except for the fact that Gustavo prefers his hair slightly disheveled, his sleeves rolled up and his alcoholic beverages to originate from grapes. He has a tasting room in the Oxbow Market in downtown Napa with a small bar up front, some tables and a view of the latest iteration of Gott's Roadside Grill across the street. "These are some amazing wines!" a woman seated with her partner calls to him across the room, to which Brambila raises his glass toward her and responds, "They're only as good as the customers who drink them." On the back wall of the tasting room, a small wooden sign reads, "Dreams come true."

WHILE MOST RESIDENTS OF the Napa Valley certainly seemed to benefit from the Judgment of Paris, not everybody celebrated it. According to one writer, "The Paris tasting had crowned a king and queen. Nothing could match the dominance of Cabernet and Chardonnay." Between 1980 and 1990,

Gustavo Brambila signs a bottle of his wine for a customer in his Oxbow tasting room. *Photo by author.*

Sign in the back of the Gustavo Wine tasting room. *Photo by author.*

the tonnage of Cabernet Sauvignon grown in Napa increased from 16,817 to 21,658, while the increase in Chardonnay was even more dramatic, multiplying from 9,060 tons in 1980 to 37,822 in 1990. The second most-grown red varietal was Pinot Noir at just over half the production of Cab Sauv, while the second most-grown white was Sauvignon Blanc at just over one-quarter of Chardonnay. Naturally, this was due in part to the fact that the king and queen commanded higher prices than any other varietals.

To those in love with other varietals, however, this was a difficult grape to swallow—especially as more and more vines were torn out only to be replaced with Cab and Chard. "We wouldn't have all this," said Bob Biale, son of Zinfandel pioneer Aldo, gesturing toward his vineyards, "if there wasn't a Judgment of Paris." He paused, looking thoughtful. "But it came at a high cost," he concluded. The vineyards to which he had gestured were planted entirely to Zinfandel. "The Napa Valley didn't used to be so bougie," lamented one longtime resident at the fact that the rising value of land after the Paris tasting had quickly priced many people out of Napa.

In 2008, *Bottle Shock*, which was filmed partly at Chateau Montelena, was released in theaters. Therein, the great Alan Rickman portrays Steven Spurrier as an unlikeable and pretentious man, so much the opposite of who Spurrier really was that the deeply offended Englishman reportedly threatened to sue over the way he was depicted. "It's absolute rubbish the way they portray me," said Spurrier. "There is hardly a word that is true in the script and many, many pure inventions as far as I am concerned." Throughout the film, Rickman repeatedly maligns California wines and Californians, at one point suggesting that bottle shock, or jet lag, might improve the wine. The character played by Rickman, however entertaining, is so unlike the real Spurrier that there can be little wonder he was irritated.

Though the movie brought further attention to the Judgment and reminded audiences of the American victory over thirty years prior, it also rewrote the story almost entirely. In *Bottle Shock* Spurrier acts alone, rather than alongside Patricia Gastaud-Gallagher, while Stag's Leap Wine Cellars is written entirely out of the script, as is Mike Grgich. Jim and Bo Barrett feature prominently, as does Gustavo Brambila. The actual Jim Barrett performs a cameo role as a winemaker.

While it appears that there was never any intention of including Warren Winiarski and Stag's Leap Wine Cellars in the movie script, actor Danny DeVito was reportedly cast to play Mike Grgich. Grgich, like Spurrier, was "shocked and dismayed" when the producers sent him the script. "I refused to sign that I approved it because so much of it was not true," stated Grgich. "It is true that the Chardonnay had won in Paris in 1976; but beyond that, I pointed out, the script contained very little that was accurate." Egregiously, the filmmakers opted to cut Grgich from the script entirely but went ahead and found an actor who didn't look much like him, stuck a dark blue beret on his head and had him lurking in the background of a few scenes at Chateau Montelena in what can only be described as a clumsy homage to *Where's Waldo?*

The collective protests of Spurrier and Grgich did nothing to cease production, and the dramaturgical disaster was released to critical acclaim. Flawed though it is, the movie accomplished two things. First, it brought greater attention and notoriety to an important historical event. Second, it helped save Chateau Montelena. In 2008, the Barrett family was in the act of selling their winery to Michel Reybier, a Frenchman, when a worldwide financial crisis struck. The sale fell through, but when the movie was released, the winery's fortunes seemed to turn. "The foot traffic increase was incredible," stated one longtime Montelena employee.

Gustavo Brambila, who in the movie plays a pivotal role in the making of the '73 Chardonnay at Montelena, was in actuality not there until 1976, the year of the Paris tasting. "The movie also portrays my good friend Gustavo Brambila as being involved in the making of the 1973 Chardonnay, but he was not," stated Grgich. Nobody, save for the film's own account, disputes this version of events. "I was there for a year and"—he shrugged—"three months?" Brambila stated casually. Understandably vexed by the film that strikes his name from the record, Grgich reported having watched it at least three times, concluding, "To this day, it bothers me greatly that people have seen and accepted the movie as the truth."

A few years later, Robert Kamen, who wrote *The Karate Kid* and *The Fifth Element*, among many others, wrote what many considered a counternarrative to *Bottle Shock*, featuring the story of Warren Winiarski and Stag's Leap Wine Cellars. Kamen had been introduced to Winiarski by George Taber, and the two developed a friendship. Kamen strongly disliked *Bottle Shock* and was intent on telling a different story. By the time the fortieth anniversary of the tasting came around, the script was written, the money was raised, a director had been selected, teasers were made and there was a buzz going around— even Spurrier was excited. The rest of the story would come to light at last. That all changed, according to Kamen, when Winiarski read the script.

"Why don't you like the script, Warren?" Kamen asked in earnest.

"You make me look like Forest Gump, like I didn't know what I was doing," responded the insulted winemaker.

"Warren, you didn't know what you were doing," insisted Kamen, which probably didn't help his cause.

The two argued. Kamen had written the script for free, and he regarded it as a "love letter" to his friend, but Winiarski wouldn't have it. The matter came to a head not long after, when Kamen was on vacation with his daughter. "I got a call in Bangkok at two o'clock in the morning from some lawyer in Los Angeles saying that 'Warren will not allow you to use his name, and if you do, he'll sue you,'" recalled Kamen. "I loved Warren, but when he pulled this stunt…" Kamen trailed off. "Warren broke my heart. He just did. I really wanted that movie made." It never happened.

Similarly, the relationship between Grgich and the Barrett family had remained strained. For years, Jim Barrett and Mike Grgich went out of their way to avoid sharing space, a difficult feat in such a small valley. Over time, however, the two softened their positions. In April 2012, as Mike Grgich was preparing to turn ninety years old, Jim Barrett emailed to wish him a happy birthday. According to Grgich, the email read:

> *In 1972, forty years ago, I had the good fortune to hire Mike Grgich as Chateau Montelena's first Winemaker. I told Mike to get the best grapes, the best winery equipment and a talented [sic] to make the wine—no expense spared. Our goal was to make world-class Chardonnay.*
>
> *Mike's reputation as a Master Winemaker was made when his 1973 Chardonnay won the "Paris Tasting" in 1976.*
>
> *Happy Birthday, Mike!*
> *—Jim Barrett*

"There was finally peace—and the truth—between us," reflected Grgich. Jim Barrett passed away less than a year later, on March 14, 2013. His son, Bo, would transition from being Montelena's third winemaker to assuming his father's role as head of the estate. On May 7, 2022, the Barrett family celebrated fifty years of stewarding their winery and vineyards. Mike Grgich, accompanied by his daughter, Violet, returned to Chateau Montelena to help the Barretts celebrate. Both Grgich Hills and Chateau Montelena continue thriving to this day.

Mike and Violet Grgich. *Courtesy Grgich Hills.*

ONCE THE MYTH OF French superiority had been dismantled, it was not only the Americans who found their way into the promised land. Indeed, the Napa Valley may have led the charge, but the doors to the cellar were open to the entire world, and winemakers around the globe were emboldened and empowered by the American victory. Soon, they began to claim victory for themselves.

In 2011, in Hong Kong, a Chilean red blend from the Aconcagua Valley known as Seña, from a winery founded by Eduardo Chadwick and Robert Mondavi—bested French First Growths Château Latour, Château Lafite Rothschild and Château Mouton Rothschild in a blind tasting. The following year, another Chilean producer, Montes, outpaced Northern Rhône wines when its 2004 Folly took first in a blind tasting in Los Angeles. In second place was a Penfolds Grange from Australia, where winemakers struggle to spell "Syrah" but nonetheless make it quite well. The French became the largest man in the bar, and every small producer with a few drinks in them was eager to pick a fight.

It is important to note, of course, that generally speaking, French wines have always been and likely always will be among the best in the world. The real gravitas of these events, these tastings in which a group of trained professionals use a scoring system to indicate that—at least at that moment—they prefer wines from other places to some of the best of France, is not to malign or detract from France's contributions to winemaking over many centuries. Instead, it's a reminder that the wines of France are not singular, they are not indomitable and when they are excellent, it is—as is true everywhere else in the world—not some mere happenstance but the result of much labor and probably also some luck. Most of all, it is to remind the world of what Chateau Montelena, Stag's Leap Wine Cellars, Smith-Madrone, Trefethen, Grgich Hills and others proved in the years following the Judgment: what makes wine great is not merely tradition and not only terroir. Above all else, it is the people who tend the vines, who crush the grapes and who guide the juice through fermentation who ultimately determine what winds up in the bottle.

Though they remained avid travelers, Steven and Bella Spurrier moved back to England, to a little town in Dorsett called Litton Cheney, to live a quiet, rural lifestyle that neither Paris nor their relative fame could have allowed. Bella had purchased some land in England in the '80s, and Steven had naturally taken note of the chalky soils. Once, on a trip to Paris, he presented Michel Bettane, a professor at the Académie du Vin, with blocks of chalk he'd gathered from their property, asking Bettane where he thought they originated.

> *"Champagne, of course," responded the professor.*
> *"Dorset," responded Spurrier.*
> *"In that case you should plant a vineyard," Bettane advised him.*

Taking the advice, the Spurriers planted thousands of vines on twenty-five acres of their estate. The Bride River flowed across their property near the house, and in gratitude for the natural beauty by which they were surrounded, the Spurriers named the estate Bride Valley Vineyard. They took advantage of the soil, continentality and diurnal ranges—in short, *terroir*—in their native England, which was comparable to that of the coveted vineyards of Champagne. In 2011, they harvested their first grapes, the Chardonnay, Pinot Noir and Pinot Meunier required to make fine sparkling wines.

In 2014, Bride Valley Vineyard released its first vintage, and soon, the small winery in Dorset had a portfolio that included a Blanc de Blancs, a Brut Reserve and a Rosé named "Bella" in honor of the woman who accompanied Spurrier throughout his storied career in wine. After an amazing life spent in the spotlight, whether intentionally or not, the Spurriers had returned home to England, bringing full circle the story of one of the most interesting and impactful lives ever devoted to the wine industry.

ELSEWHERE IN ENGLAND, SIR Peter Michael, who had opened his Knight's Valley winery shortly after the Judgment of Paris, commissioned muralist Gary Myatt to recreate the climactic moment of the tasting to display at his Vineyard Hotel in Stockcross. In an interview, Sir Peter Michael explained that the story of the Judgment is "really well known in America, it's very well known but has been suppressed in France, and is not being told at all in the UK."

"I loved the way that Sir Peter told the story," relayed Myatt, discussing his decision to take on the project. Myatt took great pains to create his version of the event, which is set around a table in an impressive stone wine cellar, as opposed to Paris's daylit Hotel InterContinental. In preparing to paint, Myatt read George Taber's book, watched *Bottle Shock* and did additional background research. The finished product is a wall covering that measures six meters in length by three meters in height, approximately twenty feet by ten feet, and took more than three months to create.

Depicted in the mural are most of those who were present that day in Paris (and one who was not). From right to left, we see George Taber with a knowing look on his face, scrawling in his notebook. Next is Jean-Claude Vrinat, engaged in conversation with Aubert de Villaine, who is holding a glass of red wine. Between them, standing up, Claude Dubois-Millot prepares to spit wine into a spittoon. To de Villaine's immediate right and behind the table stands Christian Vannequé, sipping red wine with a pinky pointed skyward, while immediately in front of him on the other side of the table sits Spurrier himself, studying his notes and perhaps avoiding eye contact with those around him.

Gary Myatt works on his mural of the Judgment of Paris at Sir Peter Michael's Vineyard Hotel in England. *Courtesy Gary Myatt.*

Right of Vannequé is Odette Kahn with a look of displeasure on her face as she consults her notes, while the expression of Pierre Brejoux, seated in front of her, is one of unmistakable horror. Behind Brejoux, an unnamed staffer in a black tie pours wine before Michel Dovaz is seen, seated, with his nose wrinkled in disgust. To his right, Pierre Tari lays his head almost on the table, looking up through his glass of wine as he seeks to get more light into the bowl. Next is Patricia Gastaud-Gallagher, extending her glass toward the black-tie-wearing staff member, apparently seeking to revisit a wine. Raymond Oliver sits at the end of the table, a look of astonishment smeared across his features as he examines the bottle of '73 Chateau Montelena.

The final person to appear in the mural is Sir Peter Michael himself, standing behind Oliver and Gallagher with a glass of wine in his hand, a nod to Tiburcio Parrott's inclusion of himself in a commissioned work. Sir Peter Michael is certainly not the first person to have himself painted into history. In fairness, Michael did attend the repeat tasting of the event held by Spurrier in 2006, while it is far less likely that Parrott ever attended a Pomo sweat ceremony.

All told, the mural is perhaps as striking to the eye as the famous tasting itself is important to the world of wine. Said Steven Spurrier of the mural: "It's the event. It's the event. It captures a dramatic moment when people are surprised, shocked, and this tells the story of the tasting." "It's a great

Mural by Gary Myatt commissioned by Sir Peter Michael for the Vineyard Hotel in Stockcross. *Courtesy Gary Myatt.*

story; it's well-told. The tasting gives way to a silence, and then an explosion, and that explosion is shown here in graphic detail," stated Sir Peter Michael fondly. "I'm really proud to put my name on it," said Myatt. Those in no way associated with the mural were equally impressed: "At the end of the corridor is a wonderful mural by Gary Myatt entitled 'Judgement of Paris'. And this praise doesn't come lightly as most hotel murals have me wondering: Why, God, why?" was the candid response of Poorna Bell from

MSN Travel. Thanks to the efforts of Sir Peter Michael and Gary Myatt, perhaps the story of the great upset will become better known in the United Kingdom over time.

IT TOOK TIME, PERHAPS more than one might have imagined, for the tremendous import of the Paris tasting and the monumental changes to the world's wine trade that followed to be acknowledged. Even with a modest budget of around $5 million, and despite not having to pay Danny DeVito to play Mike Grgich, *Bottle Shock* still somehow managed to lose money. And yet, it became what some would call a cult classic, in part because it remains one of the few Hollywood productions to focus largely on the wine trade. Thus, the real value of *Bottle Shock* may not have been as a movie but rather as the cinematic equivalent of White Zinfandel, a vehicle for introducing more people to the world of wine.

Greystone, now part of the Culinary Institute of America, long housed the Wine Hall of Fame. Brother Timothy of the Christian Brothers, who once inhabited the massive Hamden McIntyre creation, alongside Charles Krug, André Tchelistcheff and Georges de Latour, were all inducted posthumously in 2007. The following year, they were joined by Paul Draper and Mike Grgich. The year after, Warren Winiarski and *dos hermanos*—

André Tchelistcheff's plaque indicating his induction into the Wine Hall of Fame in 2007. The Wine Hall of Fame is now housed at Copia. *Photo by author.*

the Beringer brothers—joined the honored gang, along with Jack and Jamie Davies. In 2010, Zelma Long joined Jamie Davies as one of the first women inducted. Today, the Wine Hall of Fame is housed at Copia, which was established by Robert Mondavi, across the river from downtown Napa a stone's throw from Gustavo Brambila's tasting room.

Over time, greater recognition for the Judgment of Paris began to emerge, and by most accounts, the name itself, bestowed on it by George Taber, is now more closely associated with wine than it was with Greek mythology. The year 2016 marked the fortieth anniversary of the tasting. That year, the 114th Congress of the United States of America passed H.Res.734, sponsored by California Democrat Mike Thompson and cosponsored by sixty-four bipartisan representatives, "recognizing and honoring the historical significance of the 40th anniversary of the Judgment of Paris, and the impact of the California victory at the 1976 Paris Tasting on the world of wine and the United States wine industry as a whole."

That year, at the National Museum of American History in Washington, D.C., as part of the permanent archives of the Smithsonian Institution, an exhibit was created commemorating the tasting. Juxtaposed next to Julia Child's elaborate kitchen exhibit, photographs of the event taken by Bella Spurrier and two etched wineglasses from Spurrier's Académie du Vin were made part of the permanent exhibition. To the exhibit, the Barrett family donated a bottle of 1973 Chateau Montelena Chardonnay, while the Winiarskis donated a bottle of 1973 Stag's Leap Wine Cellars Cabernet Sauvignon, both full and in pristine condition. The exhibit also features Mike Grgich as the "immigrant winemaker" and displays the suitcase he used to transport his few belongings over the ocean, as well as his trademark beret. Warren Winiarski's notebook about winemaking and photographs of him and Barbara in the early days of Stag's Leap Wine Cellars are also on display in the museum, as are artifacts from Aldo Biale and Bob Trinchero's early endeavors into the world of Zinfandel.

DESPITE THE AG PRESERVE and subsequent legislation, conservation remained a major concern for Napa residents. The Napa Valley, which was once virtually unknown, became the second American Viticultural Area, or AVA, in the country in 1981. Since that time, however, the diversity of soil types and microclimates had warranted the foundation of sub-AVAs nested within Napa, and with the establishment of the Coombsville AVA in 2011, named for Napa founder Nathan Coombs, there were now sixteen nested AVAs in the valley.

In the late 1990s, the EPA declared the Napa River "impaired," in large part due to erosion and sedimentation. The Napa Valley Vintners banded together with community organizations that ultimately decided to collaborate with a program called Fish Friendly Farming to help vineyard managers and owners implement best practices to prevent erosion and soil runoff. The Napa Green Land program launched in 2004. A few years later, the vintners brought in a consultancy called ViewCraft, headed by John Garn and Anna Brittain, to develop a complementary Napa Green Winery program, which launched in 2008.

Perhaps thinking along similar lines, certainly in regard to protecting the valley, Randy Dunn sought to make even greater strides in conservation. His Howell Mountain wines had become "cult" and had made him and his wife, Lori, a lot of money. In 2013, they donated $5 million toward creating what became known as the Dunn-Wildlake Preserve, part of the Land Trust of Napa County, setting aside still more of Napa's pristine land to protect it from development.

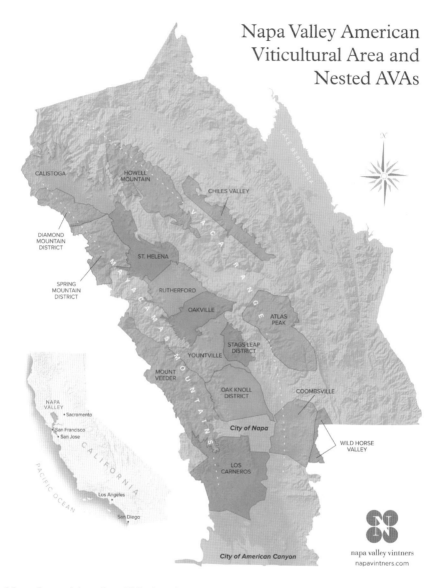

Map of nested American Viticultural Areas within the Napa Valley. Napa Valley became only the second AVA in America in 1981. *Courtesy Napa Valley Vintners.*

In 2015, the Napa Valley Vintners sought to aggressively grow their program and set a goal to have all members certified as Napa Green Land, Napa Green Winery or both, depending on their situation. Again, ViewCraft was engaged and began to grow the program and conduct resource audits

A sign indicating Napa Green Certification with Opus One in the background. *Courtesy Opus One.*

of members seeking these certifications. "From 2017 to 2019 Napa Valley Vintners and Napa Green spent significant time and resources helping growers understand and navigate the new Regional Water Board vineyard Waste Discharge Requirements. The WDR mandated erosion-control farm plans for all vineyards over five acres," recalled Anna Brittain. "Of course, environmental compliance is imperative, but it is not sustainability leadership," she added quickly.

What started as a good idea had expanded into a movement, but it was untenable for an organization primarily concerned with promoting the wine industry to serve as its own environmental watchdog. In October 2019, Napa Green split off from the Napa Valley Vintners to become its own independent nonprofit organization, with Anna Brittain as the executive director. In 2021, Napa Green announced their new, next-level Napa Green Vineyard certification (replacing Napa Green Land), setting strict standards for climate action and social equity. Some grumbling ensued, but under Brittain's leadership, the organization thrived. In the fall of 2022, the organization boasted ninety certified wineries, which included Clos du Val, Stag's Leap Wine Cellars, Grgich Hills, Larkmead, Beringer, Robert Mondavi, Schramsberg, Spottswoode, Trefethen, Opus One, Inglenook and dozens more.

The notion of sustainability was rapidly being usurped by something known as regenerative practices. It was no longer enough to sustain, reasoned many in the industry, including Brittain and Napa Green, as the status quo had already devolved to such a point that, if merely sustained, it would eventually prove catastrophic. If the world was going to survive climate change, net zero practices would need to be supplanted by something more akin to leaving things better than they had found them. But how?

For years, the idea of regenerative agriculture had been studied and experimented with around the globe, and more recently, some had begun to apply these practices specifically to the uniquely challenging realm of viticulture. Numerous practices had been identified that could, when implemented, greatly reduce the amount of carbon released into the atmosphere through traditional farming methods. Furthermore, that carbon had the potential to be returned to the soil and plant life, such as grapevines, where it could be useful. Practices like low- or no-till farming, composting and cover cropping, among others, were already being used selectively by some viticultural operations, but if greatly expanded, they could help the wine industry go beyond sustaining the status quo to regenerating the earth; increasing resilience to drought, high heat and other extremes; and actually contributing to fighting global climate change. By leading the charge in regenerative viticultural practice, the Napa Valley could put its influence to use for the greater good. But even if the entire area made a full transition to regenerative practices, the world's climate had changed. Napa was hotter and drier than ever before, and those conditions made the entire valley, from Mount Saint Helena to the San Pablo Bay, a fire hazard.

FIREBELLE LIL, AS SHE was affectionately referred to, had made Larkmead her home, whether by her own choices or those of her parents, who sought to hide her seemingly unslakable thirst for flouting social norms from the public eye. But long after Lillie Hitchcock Coit returned to San Francisco to live out her days in the company of the firemen she idolized and supported, the Larkmead property she had named and inhabited continued producing wine. The property changed hands several times, eventually coming into the possession of Cam and Kate Solari-Baker. The couple had much of the property replanted and soon began renovations and additions to the buildings.

Fire was nothing new to the Napa Valley. Meticulous stewards of the land, the Wappo had been doing controlled burns since long before the arrival of the Spanish. The Wappo coppiced shrubs and burned significant swaths of land, removing fuel from the path of unplanned fires while eliminating the stunted flora that interfered with hunting for game. Furthermore, the Wappo knew that some species of plants rely on burning to stimulate sprouting and growth. They understood fire to be a tool in a way that those who followed them—or rather, ran them out—never managed to fully comprehend.

Since the time of the Wappo, fire remained a near-constant companion to the residents of the Napa Valley and surrounding areas. An 1870 headline in the *Sonoma Democrat* bestowed the moniker Great Fire on three separate blazes, two in St. Helena and one in Calistoga, that converged to sweep across the Mayacamas. In September 1913, a fire originating in the Chiles Valley was deemed "the worst fire in the history of Napa County" by the *Saint Helena Star*. In 1931, a fire originating in Berryessa

Valley torched forty thousand acres and was called "one of the worst in the history of Napa County."

Around the start of harvest in 1964, three separate fires consuming more than seventy-five thousand acres prompted the *Register* to run the headline "Valley Is Flaming Nightmare," and the following year, eight simultaneous fires burned an additional forty thousand acres around the town of Napa. Raising flammable crops in a rural area where the steady breeze from the San Pablo Bay was liable to fan flames and spread cinders through the air was, in some ways, the accepted lot of Napans. Fire was a part of life; there was no getting used to it and no getting rid of it, but most believed that it was possible to fight it and that the benefits outweighed the risks.

Then, in October 2017, it seemed as if the entirety of the Napa Valley was set ablaze. In what many termed a repeat of the 1964 fires, "though on a larger scale," three separate blazes encircled the valley, filling it with smoke and threatening the lives of every living thing therein. The Tubbs Fire consumed thirty-seven thousand acres. Down valley and east of the town of Napa, the Atlas Fire devoured another fifty-two thousand acres, while to the west in the Mayacamas, the Nuns Fire blazed through another fifty-six thousand acres. In addition to ancient redwood groves, wildlife habitat and vineyards, these fires destroyed 650 homes. Wine country had survived phylloxera and the Prohibitionists; now, more severe fires than ever before threatened to wipe it off the map.

The year 2018 saw what was, at that time, the deadliest and most destructive fire season in the history of California. More than eight thousand individual blazes consumed nearly 2 percent of the state's surface area, killing more than one hundred people and burning over twenty thousand structures. Despite these staggering figures, Napa enjoyed a brief respite from the blazes as many attempted to rebuild after the devastation of the previous year. In 2019, the Kincade Fire in adjacent Sonoma forced two hundred thousand people—nearly twice the entire population of Napa County—to evacuate their homes and caused many in Napa to lose sleep at the prospect of embers catching a Pacific breeze and spreading through the Mayacamas.

Then, in 2020, when the COVID-19 pandemic forced millions indoors and ground wine country's tourist industry to a halt, fires again struck the Napa Valley, more severe than ever before. Caused by lightning strikes and fueled by overly dense, parched forests and agricultural land, the LNU Lightning Complex Fires were among the largest and deadliest in the history of the entire state. Many Napa residents evacuated. Others hunkered down

The iconic Napa Valley sign, with the mountains behind it ablaze. *Photograph by Tim Carl.*

at their wineries, determined to defend their livelihoods or else be devoured with them. All told, the LNU Lightning Complex Fires of 2020 would burn for a month and a half, devouring 165,000 acres, nearly one-third of Napa County, while killing six people and sending hundreds of houses and several wineries up in smoke.

As if the flames themselves were not enough, it turned out that fire and burning weren't the only threat to the grapes. Where there's smoke, there's fire, and fire's unmistakable herald proved to be just as effective at ruining the crop as the flames that could consume it. The term *smoke taint*—in essence, the flavor of smoke on grapes being transferred into wine—began to spread like wildfire and inspired nearly as much alarm. While many of the white grapes had been harvested by the time the LNU Lightning Complex Fires began ravaging the valley, most of the red grapes, which tend to mature later, remained on the vines and were soon enveloped in a thick blanket of ashy, blackened fumes.

Down in Oakville, on the valley floor, Opus One managed to harvest a little bit of fruit before the smoke set in, but not much. "It will be our second smallest vintage ever," said winemaker Michael Silacci. In the winery's first vintage back in 1979, it made just 2,500 cases. Since that time, Opus One's production had increased annually until it reached nearly 30,000 cases per annum. In 2020, however, Opus would produce only 3,100, approximately 10 percent of its usual production. Their nearby Oakville neighbor, Dennis Groth, also reported planning "minimal production" of the 2020 vintage.

Over in Stag's Leap, Bernard Portet sighed. "The smoke settled on the vineyards. There was no wind to push it out." Clos du Val would be among the many wineries not to make a 2020 vintage of red wine at all. Not far away, Gustavo Brambila smiled his usual patient smile. "No, I didn't make a 2020 vintage due to the smoke," he said. Despite Brambila's access to choice vineyards in the Coombsville, Mount Veeder, Atlas Peak and Diamond Mountain AVAs, his exacting standards had a tendency to get in the way of profit. "I didn't make a 2017 either," he shrugged.

From Smith-Madrone on Spring Mountain to Chateau Montelena near Calistoga, the story became increasingly familiar until it grew repetitive. More and more vintners were choosing to dump their reds in 2020 rather than put their label on a product they felt was inferior. Dump them if, of course, the vineyards hadn't been lost to the flames. If, of course, the wineries hadn't been burned to the ground. While roads and rivers offered an added degree of insulation to those on the valley floor, wineries and vineyards in the mountains were significantly less protected, and many were set ablaze.

Spring Mountain Vineyards was hit hard by fire in 2020. In addition to ninety-six thousand vines lost to fire, nearly a dozen nineteenth-century buildings on the property were also consumed in flames. While Miravalle and several adjacent buildings were saved, the historic three-story La Perla Winery burned to its foundation, as did the house at the top of the mountain where Jerome Draper had stored duplicates of the wines tasted at the Judgment of Paris for future tastings. The historic Chateau Chevalier was licked by flames that came within twelve inches of it, consuming the densely overgrown trees and brush that surrounded it, though ultimately the abandoned chateau was spared.

It would take time, as well as bullheaded determination, for the vintners of Napa to rebuild and replant again, but most were determined to do precisely that. They knew that, however incendiary it might be, their valley possessed not only unparalleled beauty but also equally unparalleled capacity, with its exceedingly diverse soil types and myriad microclimates, to produce world-class wine. And then, from the smoldering ashes of ruin, there emerged yet another difficulty.

Somewhere in the unctuous, miasmic bowels of corporate America, the number crunchers were losing their minds. A planted acre of vineyard in Napa could easily be valued at half a million dollars—five hundred times that of land in other parts of the country—and the damn things were going up in smoke forty thousand at a time! Housing costs were on the rise nationwide, but the median home price in Napa was approaching $1 million, and those were just as—if not more—flammable than vineyards. And in such a limited geographical footprint, even the modest, inornate wineries were worth millions. With these staggering figures illuminated in the firelight, the insurance companies were beginning to conclude that there was no amount of money that could suffice as a premium to insure the Garden of Eden against the raging fires of hell. Many insurers made doubling premiums an annual tradition as predictable as the winter solstice. At that rate, it didn't take long for small wineries to balk at the skyrocketing expense of insuring themselves against the flames.

Doorhanger indicating the structure has been evacuated due to fire in St. Helena. *Courtesy Geoff Ellsworth.*

"It's a lot more than it used to be," said Violet Grgich of the premiums that Grgich Hills was paying. "We still have insurance, but…" she trailed off, raising her eyebrows in place of finishing the sentence. "It's six times what it was five years ago," sighed the usually cheerful Dennis Groth in frustration. "The hillsides are almost uninsurable now," said Nils Venge, Groth's former winemaker, contemplatively staring up into the Vaca Range. Up and down the Napa River, from the valley floor into the mountains, the story was the same. The response of the insurance industry to the wildfires was, in essence, what it had always been: charge enough money in premiums that even when they had to pony up for damages, they'd still come out ahead.

"Tired," responded Stu Smith, when asked how he was doing, and he sounded like it. In late August 2022, he and Charlie were in the midst of harvesting their coveted Riesling. Napa had once again been scorched and parched, and in addition to the usual toil of harvesting grapes, Stu had other things on his mind, to boot.

"I feel like Satchel Paige," said Stu, quoting the greatest pitcher of all time. "Don't look back. Something might be gaining on you." In the fall of 2022, a lot of things might have been gaining on Stu and Charlie, the same things that were gaining on so many other small producers. "The insurance company just doubled our rates," Stu explained, adding that in 2022, Smith-Madrone would pay more than $100,000 in fire insurance alone. "In 2019, we paid $16,000," he said, punctuating the point that between the fires and the insurance companies, what was gaining on the small wineries had the

Vineyards threatened by fire at Smith-Madrone. *Courtesy Smith-Madrone.*

potential to catch and devour them. "I'm considering going naked," Stu went on, using a common term for taking the risk of operating a winery in the ever-incendiary valley without insurance. He was far from the only one threatening to do so—so far, in fact, that if something wasn't done soon to make fire insurance more affordable, the Napa Valley was liable to become a nude beach.

Down Spring Mountain from Smith-Madrone, Keith Baker at Spring Mountain Vineyard hung the radiator from his truck next to the entrance to the tasting room. Having melted in the fire that torched much of the estate, it now resembled modern art far more than part of an automobile. But while the remnant of radiator looked kind of cool, the scorched hillsides where Spring Mountain's vineyards had once grown and the still-smoldering foundations of so many historic buildings that were part of their estate were anything but. The staff at Spring Mountain worked diligently to rebuild, but it was an uphill battle.

All told, 2022 would mark what many termed a five-hundred-year drought, indicating the most severe drought conditions the area had experienced in half a millennium. On Thursday, September 1, 2022, the *Napa Valley Register* released its harvest report. The title: "Here Comes the Heat." On September 7, the *New York Times*'s daily email read, "The Morning: A Summer of Climate Disasters," while CNN's competing publication was titled "A Record Setting Heatwave Is Scorching the Western US" and featured an image of a man holding ice to his head and eating a popsicle in Santa Rosa. The very next day, September 8, the front page of the *Register* read, "Napa Looks at Its Future in Wine Country" and displayed a large image of Lake Berryessa with water levels so low the reservoir looked more like a pond. That day, the high in the town of Napa was 104, while up valley Yountville reached 107, St. Helena 108 and Calistoga 109. When the *Register* sent out its weekly harvest report that afternoon, this time the title was "The Heat Is On." No kidding. On Monday the nineteenth, the previous day's rainfall, which totaled less than an inch across the Bay Area and, in parts of Napa, less than one-quarter, made front page news. "Rain Brings Relief," touted the *Register*. Somewhere nearby, Stu Smith sighed.

At Larkmead, where Firebelle Lil had once whiled away her time in the peaceful, rural absence of San Francisco, the extreme warmth of the northern end of the Napa Valley became a testing ground for viticultural experiments. Dan Petroski, Larkmead's longtime winemaker under the Solari-Baker ownership, busily planted an array of varietals that conventional wisdom suggested might struggle. The intent was in large part to determine a path forward given the looming realities of global climate change in Napa. In 2021, Petroski left Larkmead, soon after becoming a board member at Napa Green, and Firebelle Lil's old stomping grounds were again overseen by a woman when Avery Heelan, previously the assistant winemaker, assumed the reins. The people of San Francisco had named a towering monument Coit Tower in honor of Firebelle Lil's generosity toward the fire department

How the Judgment of Paris Put California Wine on the Map

that once saved her life and made her an honorary member. Somewhere out there, still wearing overalls and puffing on a cigar, Lillie Hitchcock Coit took stock of her old home, saw Avery Heelan in charge of the winemaking and raised a glass to her.

It was true enough that, in some isolated instances, women like Josephine Tychson, and later MaryAnn Graf and Zelma Long, had begun to forge a pathway for women in the wine industry. It was equally true, however, that as recently as a century ago, convention held that a woman's place was in the home, making food and babies, and the same antiquated, backward worldviews that had banned alcohol had seen to it that being a working woman in America was a tough row to hoe. Meanwhile, the protests following the murder of George Floyd awakened in the Napa Valley what some considered a long-dormant consciousness that this place of incomparable beauty and immeasurable wealth perhaps had obligations beyond the fermentation of grape juice.

"Sometimes, I'm in places, and they look like places I should not be." Victoria Coleman chose her words carefully. Dark-skinned, with eyes that are as intelligent as they are penetrating, the only Black woman making wine in Napa endures a raft of difficulties that others in wine country do not. "I don't feel safe sometimes in certain places," said Coleman. Even after decades in the Napa Valley, Coleman stands out as a Black woman in wine country. Did she expect that over time she would be joined by more people who looked like her? "I can't say that, no," she responded. "Maybe like one or two, but not…" she trailed off.

Coleman's family thought that wine, and Napa, were odd choices for her. "My family's very proud of what I've accomplished, but I think they thought it was weird at the same time," she reflected. Coleman originally found herself studying computer science in Seattle, but when her mother died, she felt she needed to get out. She chose California, and Napa, and the wine industry found her. "My parents did drink wine as I was growing up, but I never really thought about it or where it came from," recalled Coleman. Today, Coleman is where the wine comes from, and it is fantastic wine indeed.

But when George Floyd was murdered in 2020, Coleman was reminded yet again that she was in a place where the melanin in her skin made her stand out in no small way. The microaggressions, such as people assuming she worked at Brown Estate, one of the few Black-owned wineries in Napa, turned macro, like people yelling things at or around her in a local bar. "You don't know where you fit in sometimes," said Coleman with a slight shudder.

In the summer of 2020, Coleman found herself with a friend in nearby Boonville, an unincorporated and decidedly rural community north of San Francisco. They were there to pick apples. All of a sudden, in the tiny town, people began to gather. "I could see this protest getting ready to happen," recalled Coleman, whose pulse quickened as agitated people who did not look like her began to gather. Then she read their signs, heard their chants and felt something akin to love. "We were in the middle of nowhere where I might have felt unsafe, and here are these people in this little town in Anderson Valley marching in support of Black Lives." A gentle smile found its way to her previously reticent expression.

Disparities within the wine industry were further magnified when an exposé in the *New York Times* revealed what some were painfully all too aware of: far too many women had endured unthinkable abuse in their quest to become sommeliers. The roots of the problem ran deep in the Court of Master Sommeliers, Americas. Years before, a documentary titled *Somm* had vaulted the profession into the spotlight. "Suddenly, sommeliers were celebrities," recalled Emily Wines, one of the rare woman master

Hundreds of demonstrators gather in support of Black lives in downtown Napa on May 31, 2020. *Photo credit:* Napa Valley Register.

sommeliers. Seemingly as a result of the film, enrollment in entry-level classes skyrocketed with the newfound recognition that being a sommelier was not only viable but also a sexy career path.

"I remember I was in Napa, working here during the Napa Valley Film Festival in 2012, when that documentary came out, and it was just electric," remembers Vincent Morrow, one of four African American master sommeliers in the world and the chairperson of the court's diversity committee. "The energy was palpable," he recalled. "I feel like that was the beginning of a meteoric rise for the profession."

But the swell of newfound interest in joining the industry on the side of service created unforeseen problems for the court. In partial response to sharp increases in enrollment and the demand for testing far outpacing the court's ability to administer the requisite exams, a policy was introduced requiring advanced sommeliers to have the recommendation of a master. "It created a power imbalance," reflected Wines.

"There were really three crises," states Julie Cohen Theobald, who was made executive director of the Court of Master Sommeliers, Americas, in September 2021 as a direct result of the crises she was recalling. "The first was in 2018, and the invalidation of an exam due to a breach of security by someone on the board. It shook the confidence and integrity of the organization," she says, referring to a widely publicized cheating scandal. "The second crisis was the George Floyd slaying. The organization didn't take a stand. We really made some mistakes and didn't have the commitment to diversity that we needed," stated a pensive Theobald. "The third crisis was the *New York Times* article… There was a petition that thousands of people signed saying that the organization should be dismantled," she recalled.

In 2020, at the tail end of the #MeToo Movement, the upper echelons of leadership in the Court of Master Sommeliers, Americas, was toppled from the inside out. What had been known only to a small handful of women and perhaps their closest confidants was brought to light in the *New York Times*, and the veil was lifted. Some of the biggest names in wine, some of the most powerful, influential men in the industry, were accused of being sexual predators or even worse. Shortly thereafter, the entire board of directors resigned. Four women, all master sommeliers, left the organization in protest. "It was devastating," recalled Wines. "The reporting line was flooded." More and more master sommeliers—all men—were suspended. An outside organization, the same one used by the NFL, was hired to investigate the problem. "There were calls for blood," recounted Wines.

"I knew some of the women," said Zelma Long. "It's no different than what's happened in many industries. To me, it's just another example of how men take advantage of women."

Though in the eyes of some, the damage was already done, the Court of Master Sommeliers, Americas, launched into correcting the problem. "We radically overhauled our code of ethics," stated Wines, who was also instrumental in the hiring of Julie Cohen Theobald as the organization's executive director. The court permanently banned many men who had allegedly been among the worst perpetrators, formally stripping them of their title as master sommelier. The court also began making a far more concerted effort to be inclusive. They began partnering with historically Black colleges and universities to bring more people of color into the profession. "I'm really excited about the possibility of it," said Morrow. "The pace of change has exceeded my expectations," said Theobald proudly, "and I came in to make change."

Today, Emily Wines is the chair of the board of directors for the Court, and Kathryn Morgan, a fellow master sommelier, is the vice chair. Of the other nine at-large positions on the board, only one other, Mia Van de Water, is a woman. In total, there are twenty-five women master sommeliers who are members of the Court of Master Sommeliers, Americas. Wines called the figure "deplorably low." But the gender bias in the wine industry isn't limited to the service side of things, of course. According to Bâtonnage Forum, whose stated purpose is to "Stir It Up," only 14 percent of the thousands of wineries in California employ a woman as their head winemaker. Furthermore, Bâtonnage Forum reports that the median annual wage discrepancy between male and female sommeliers is a whopping $7,000.

The court wasn't the only group to take note of the need and step up in an attempt to diversify the industry. Within months of the murder of George Floyd, the Napa Valley Vintners joined forces with the United Negro College Fund by pledging $1 million in scholarships to Black and Indigenous students seeking to enter the wine industry. After the first year of the initiative, however, only around $19,000 in scholarship money had been claimed. The Napa Valley Vintners asserted that this was due to a lack of applicants. The diversification of the wine industry would not happen overnight.

VICTORIA COLEMAN'S FIRST JOB in Napa was at Stag's Leap Wine Cellars. "I was supposed to be there two weeks," laughed Coleman, who initially took the job to fill in for an employee who was getting married. She ended up staying, learning all aspects of the business and working in every part of it, as well. "I felt most comfortable in the vineyard," reflected Coleman. "She's very sensitive to wine," remembered Warren Winiarski. "She did her job at the winery very well. I'm very fond of her," he added quietly.

Coleman remembers her time at Stag's Leap Wine Cellars fondly and the man who hired her just as much so. "He was just always so philosophical about wines, it was almost scary," laughed Coleman, who also recalled that Winiarski loved to dance. "He used to take lessons in the white-wine-making building," she remembered. "Warren still adores her," relayed Opus One's winemaker, Michael Silacci, who also worked for Winiarski when Coleman was hired. Eventually, Coleman left Stag's Leap Wine Cellars to study at UC Davis. Today, Coleman is the winemaker at Lobo Wines. Progress is arguably being made, in the wine industry and elsewhere, but that progress is far too slow.

Other wineries that had been part of the Judgment of Paris were also faring better than the national average when it came to employing women in their industry. Freemark Abbey, where Josephine Tychson had become the first woman winemaker in California a century and a half before, hired Kristy Melton as head winemaker in 2020. Melton had previously been making wine with Bernard Portet at Clos du Val, and on her departure, Clos du Val hired another woman, Carmel Greenberg, a native of Tel Aviv, who had previously made the wine at Napa's Dominus Estate. At Heitz, Brittany Sherwood serves as director of winemaking, while Hannah Hercher works as the oenologist.

Others, like Chateau Montelena, still had men at the helm yet made a concerted effort to employ more women, such as oenologist Jamie Eggerss. Heidi Peterson Barrett, Bo Barrett's wife, had become one of the most sought-after winemakers in all of the Napa Valley, referred to as a "wine goddess" by some. Barrett had her own label, La Sirena, and also made the wine for a litany of high-end clients, among them, Ren Harris and Jean Phillips, former partners in real estate. Phillips started her winery, Screaming Eagle, in 1992, and Harris started his, Paradigm, with the release of the 1991 vintage, both of them in Oakville.

The Barretts' daughter, Chelsea, possesses talents similar to those of her parents. Currently, Chelsea Barrett serves as the winemaker at Napa's Materra, owned by the Cunat family, where another woman, Caryn Harrison, is her assistant and a third, Maggie Purdie, is the enologist.

While Paul Draper continues making the wine at Monte Bello, Ridge hired Shauna Rosenblum to make the wine at their Lytton Springs property

Left to right: Bernard Portet; Olav Goelet, grandson of John Goelet; and Carmel Greenberg walk the vineyards at Clos du Val. *Courtesy Clos du Val.*

The entrance to the David Bruce Winery. *Photo by author.*

in Sonoma. With the passing of David Bruce in 2021, his widow, Jeannette Bruce, became the face of their winery. "My vision is to keep the winery going and make the greatest Pinot Noir ever," she said. When Mike Grgich retired, it was his daughter, Violet, who took the reins at Grgich Hills in her capable hands. The white-male-dominated world of wine was ever so gradually becoming less so, though at the current rate of change, anything akin to true equity in Eden seemed unlikely to arrive prior to the rapture.

The Napa Valley exploded into a period of unprecedented growth when a wine made by a Croatian who studied under a Russian who was brought to the United States by a Frenchman shattered a myth and paved the way for the globalization of the modern wine industry. Yet even after the Washington football team and the Cleveland baseball team begrudgingly changed their dehumanizing mascots, the remaining Wappo—who had long ago departed the Napa Valley—were no closer to regaining a foothold in the fertile breadbasket of their inheritance than they were under the rule of the Spanish. The long arc of history may indeed bend toward justice, though ultimately a great deal more is required if the industry is to achieve equity and the much-needed representation of women, Hispanics, African Americans and others in the homeland of the Wappo.

Today, up and down the Silverado Trail and Highway 29, the two parallel arteries that connect the vineyards, wineries and hamlets that rest in the valley between the Mayacamas and the Vacas, travelers and locals alike can catch fleeting glimpses of the rich history of the area that surrounds them. Betwixt the ghosts of George Yount's Rancho Caymus and Edward Bale's Rancho Carne Humana, the great arch that was built by Robert Mondavi after he parted ways with his family still stands sentry on the western side of Highway 29, the same road he and his brother once traversed with their father, Cesare, to scout out the then-defunct Charles Krug Winery, which also still stands a little farther north, tended to by the dedicated descendants of Robert's brother, Peter. The brothers finally reconciled in 2005. Robert died three years later, and Peter passed away in 2016.

Farther south on the valley floor, in what was once Rutherford, a small town named for the man who married George Yount's daughter, the old Ewer and Atkinson Winery that Georges de Latour's wife, Fernande, had named with her French utterance of *"Quel beau lieu!"* remains. This is Beaulieu, where the Maestro himself, a Russian refugee who perhaps did more to shape the world of wine than any other single individual in history,

The statue of Tchelistcheff toasting a visitor. *Photo courtesy the author.*

once worked before being driven mad—and driven out—by corporate greed and ineptitude. Beaulieu was purchased from UK-based Diageo in 2015 by Treasury Wine Estates, which also owns Beringer and several dozen other wineries around the world. Here, on a shaded cement patio adorned with simple chairs and tables and surrounded by hedges that muffle the sounds of nearby traffic, André Tchelistcheff can still be found in the form of a life-size statue of the small man who made such an immense impact, his wineglass outstretched in a toast to all who seek him out in this beautiful place.

In 2022, the Calistoga winery founded by rope magnate Alfred Tubbs just east of José de los Santos Berryessa's Rancho Mallacomes 140 years prior celebrated 50 years of Barrett ownership. Bo Barrett was now at the helm, and Matt Crafton had taken over winemaking. Chateau Montelena continues to this day to produce one of the most recognizable and noteworthy Chardonnays on the market, along with phenomenal Cabernet Sauvignon and other varietals. Asked at a gathering at Chateau Montelena if he thought one day he'd be signing books about the Judgment of Paris, George Taber laughed. "No—no idea!" Today, there are eleven bottles of Grgich's '73 Chardonnay still preserved at Montelena. Jim Barrett's visionary gambit lives on, his legacy secure, his name indelibly etched into the storied history of the prestigious wine region that he and others helped put back on the map after Prohibition.

Traveling south past Edward Turner Bale's impressive mill, which is still in operation, the winery where Josephine Tychson became the first woman winemaker in Napa if not the entire United States, today called Freemark Abbey, continues to welcome visitors. Inside the beautiful tasting room, stone

Left to right: Warren Winiarski, George Taber and Mike Grgich. *Courtesy Grgich Hills.*

and wood from Tychson's original cellars have been repurposed into the architecture, which is as warm and inviting as the wines of Kristy Melton, the woman who makes the wine at Napa's first woman-owned winery.

Continuing south to St. Helena, Spring Mountain Vineyard and the beautiful Miravalle built by the outlandish Tiburcio Parrott still stands despite nature's persistent efforts to burn the property to the ground. The Glass Fire, which took out eleven historic buildings on the property, left its mark in more ways than one. Under the ownership of Jacqui Safra, the quiet Swiss billionaire, the estate filed for Chapter 11 bankruptcy in October 2022 while also suing its insurance companies for claims against the damages caused by the fires. The set of *Falcon Crest*, where the Beringer brothers used to pass time with Parrott during the pre-Prohibition era of the Napa Valley, remains strikingly beautiful, its wines as profound as its future is uncertain.

Farther south still, up the winding switchbacks of Lokoya Road, John Henry Fisher's Mayacamas Vineyards gazes down on the valley floor below. Today, the extinct volcano where Fisher first built his isolated winery welcomes visitors and produces some of the most noteworthy and coveted expressions of Cabernet Sauvignon and Chardonnay in all the world. And for those whose stomachs may not be up to the dizzying heights of ascending Mount Veeder, the year after the Schottensteins assumed full ownership, they opened a Mayacamas tasting room on First Street down in the town of Napa, as well.

In 2018, Heitz Cellars was sold to Gaylon Lawrence Jr., who made his money owning farmland and banks. "When we met with Gaylon, it seemed a perfect match. In the wine business we are all farmers, and with

Right: Freemark Abbey winemaker Kristy Melton. *Courtesy the Jackson family.*

Below: Ancient aging vessels and a comparatively young pup at Mayacamas. *Photo by author.*

the Lawrence family's history in agriculture, we feel Heitz Cellars will be in good hands," said Kathleen Heitz-Myers at the time. Not long after he acquired the property, however, Lawrence Jr. had rendered Joe Heitz's legacy all but unrecognizable, changing the iconic label designed by his son and transforming the rural-feeling tasting room that had, until recently, been one of the few remaining holdouts to offer free tastings in the Napa Valley into a decidedly un-agrarian celebration of opulence in white tablecloths and crystal, surrounded by manicured hedges and proudly advertising $1,000 tastings. If one listens closely, one can still hear the charming curmudgeon muttering to himself from the grave whenever visitors to the tasting room that bears his name are affronted by the exorbitant expense and Fifth Avenue atmosphere of his once-so-welcoming winery.

Over on the Silverado Trail, near José Ygnacio Berryessa's Rancho Chimiles in what today is the Stag's Leap AVA, Warren Winiarski's Stag's Leap Wine Cellars remains under the management of Ste. Michelle Wine Estates. Today, the FAY Outlook and Visitors Center offers striking views of the vineyard that produced the wine that first captured Winiarski's imagination and inspired him to make the Cabernet Sauvignon that would forever improve the fortunes of the wine industry in the Napa Valley. Winiarski still grows grapes, tending the Arcadia Vineyard that he purchased from Austin Hills a few decades prior. Mike Grgich once tended to this vineyard. "I'm using his hints about this place for the Chardonnay," said Winiarski, who sells the grapes he grows back to Stag's Leap Wine Cellars, where Marcus Notaro makes them into wine. Asked about the history of the winery and Winiarski's success, Notaro chuckled and recounted, "My first harvest I was here, there was this couple that was lost." After Notaro helped them find the tasting room, they congratulated him. "For what?" asked a confused Notaro. "Back in 1976, you won the Paris tasting," the couple responded enthusiastically. "Congratulations!"

Only a mile to the south as the crow flies, still in the Stag's Leap AVA, Clos du Val continues greeting visitors, as well. Bernard Portet, who first selected the site nearly fifty years before, remains a fixture in the facility, though he has enthusiastically relinquished his winemaking responsibilities, and today, Carmel Greenberg crafts the wines for Clos du Val. "Really, I'm the past," says Portet modestly. "What excites me is the future." Still, one can't help but marvel when they see the still-spry Portet walk through the door of the tasting room nearly fifty years after he first arrived.

In *Somm 3*, the third installment of the cult favorite documentary series on sommeliers and wine, which was notably released two years before the

Above: The entrance to Clos du Val winery, celebrating fifty years in Napa. *Photo by author.*

Opposite: Steven Spurrier strolls beneath the signpost denoting Steven Spurrier Lane in front of Clos du Val. *Courtesy Jason Wise.*

allegedly predatory nature of so many in the court came to light, Steven Spurrier quipped that he didn't want the Silverado Trail, but he wouldn't mind having a little street named for him. Of note, Spurrier and his companions were drinking the '72 Clos du Val when he said this. Whether by coincidence or—more likely—not, the road leading from the Silverado Trail to Clos du Val winery was renamed Steven Spurrier Lane in 2018, the same year that *Somm 3* was released. Shortly before his passing, in a 2021 interview, Spurrier admitted to having planted this idea. "Steven Spurrier Lane—it's exactly up my street," quipped the convivial Briton.

Steven Spurrier died on March 9, 2021. "We played 'Ain't Misbehavin' as he was carried out of the church," recalled Bella Spurrier. Not long after, an eightieth birthday celebration was held for Spurrier posthumously in Napa. "Angela Duerr organized it, her and [Jean-Charles] Boisset," recounted Bella, who did not attend as she was still in mourning and didn't want to bring the mood of the celebration down. In attendance were Patricia Gastaud-Gallagher, Paul Draper and countless others from around the industry and around the world, all of whom sought to pay their respects to the man who changed the world of wine forever.

Paul Draper still spends his days high up on the gorgeous mountaintop where Ridge Vineyards remains at the pinnacle. Sharp and spry, Draper is the constant presence behind the incredible consistency of Ridge Monte Bello. In the early morning, a thick blanket of fog often covers the valley floor below, and visitors who drive the winding switchbacks to the crest of Monte Bello weave their way precariously through the clouds, emerging high above them in the sunlight with a view of San Francisco in the distance. Ridge continues to produce Monte Bello Cabernet Sauvignon, and the wine tasted at the Judgment of Paris that won many subsequent tastings has become a sought-after bottle among the most discerning wine lovers around the world.

Elsewhere in the Santa Cruz AVA on the mountainsides above the Silicon Valley, David Bruce Winery has also endured the test of time. While Bruce passed away in 2021, his wife stands steadfastly at the helm, a bottle of the Chardonnay tasted at the Judgment of Paris resting on her desk. The year of Bruce's death, the winery nearly burned to the ground, and they lost many vineyards to the fires. Smelling blood, the sharks began to circle. One would-be buyer showed up with an open checkbook. "'Name your price,' he told me," recalled Jeannette Bruce. "I wasn't interested." Before David's death, he and Jeannette had set in motion their own vision for the future, and in 2022, the David Bruce Winery entered into an unlikely partnership with Hillsdale College, a private liberal arts college in, of all places, Michigan. "They match our values," said Jeannette Bruce. "My vision is to keep the winery going," she added, "and to make the greatest Pinot Noir ever."

Farther south still, in the shadow of the impressive Pinnacles with a view of Big Sur far off in the distance, Chalone Vineyard, too, continues to thrive. East of the Salinas Valley that's so often blanketed in fog, this is a rare and isolated place where the endangered California condors still soar gracefully overhead. Even with the state's untenable population growth, Chalone feels exceedingly rural and off the beaten track, and the conspicuous absence of "Chateau" draws a clear line between this largely quiescent setting and the modern Napa Valley. What had been deemed a national monument by President Roosevelt in 1908 was made a national park in 2013 by President Obama, and with the passing of that legislation, the road to the winery was paved for the first time. Chalone remains a place few stumble upon by accident but to which many make their pilgrimage.

Halfway around the world, on an Adriatic peninsula in his native Croatia, Mike Grgich's Grgić Vina Winery welcomes visitors, sharing wines made from Plavac Mali and Posip, native Croatian varietals, as well as an

Sign welcoming all to Ridge Vineyards. *Photo by author.*

assortment of the wines made in Grgich's winery in the Napa Valley. "It is a good thing to be recognized for what you have achieved," reflected Grgich, "but sometimes I think the greatest honor comes from the people who drink and enjoy my wines."

Back in Napa, Zelma Long came to visit her mentor in October 2022. They spoke of his work in Croatia, her work in South Africa and how far the

Grgić Vina, Mike Grgich's winery in his native Croatia. *Photo by author.*

American and Croatian flags fly at the entrance to the Grgich Hills tasting room. *Photo by author.*

The fairest one. *Photo by Chuck O'Rear.*

industry had come since they first met. Before leaving, she thanked him for everything he did for her. In the spring of 2023, preparations began to help Miljenko "Mike" Grgich, the immigrant winemaker who changed the fate of the Napa Valley and the world, celebrate his hundredth birthday, marking a century of life well lived.

THIS MORNING, THE NAPA River flowed gently southward from its origin on Mount Saint Helena toward the tranquil waters of the San Pablo Bay, and the sun emerged from behind the Vaca Range, warming the valley and burning off the fog as it passed in an arc, on its way to disappear again behind the densely forested Mayacamas at the end of the day, just as it has always done. What lies between these points of interest has changed hands and forms countless times over thousands of years, though the fact that it is a place of seductive and singular beauty, devoted to agriculture and filled with skilled craftsmen and those who love the land, remains as true today as it has always been. In the end, no matter how long Bordeaux had been there, no matter how storied the soils of Bourgogne had become, no matter the *what* or the *where*, the *who* or the *when*, in the end, there could be no question. Paris had spoken. The Napa Valley had always been—and forever will remain—the fairest one.

SELECTED BIBLIOGRAPHY

Chapter 1

Almaden Vineyards. "About Us." Accessed August 10, 2018. https://www.almaden.com/about-us/.

Bacich, Damian. "Governors of Alta California." The California Frontier Project. September 6, 2017. https://www.californiafrontier.net/governors-alta-california/.

Brown, Alexandria. *Hidden History of Napa Valley*. Charleston, SC: The History Press, 2019.

Chalone Vineyard. "Our History." Accessed November 23, 2021. https://www.chalonevineyard.com/About/Our-History.

Chateau Montelena Winery. "The Chateau." Accessed November 16, 2021. https://web.archive.org/web/20180921104351/https:/montelena.com/winery/the-chateau.

————. "History." Accessed November 16, 2021. https://web.archive.org/web/20180921025533/https:/montelena.com/winery/history.

Chroman, Nathan. *The Treasury of American Wines*. New York: Rutledge Books, 1976.

City of Napa. "History of Napa." Accessed December 9, 2021. https://www.cityofnapa.org/282/History-of-Napa.

Conaway, James. *Napa: The Story of an American Eden*. Columbia: Houghton Mifflin Harcourt/Open Road Integrated Media, 2002.

CPN Cultural Heritage Center. "History." Accessed November 19, 2021. https://www.potawatomiheritage.com/history/.

Far Niente Winery. "Our Story." Accessed January 16, 2022. https://farniente.com/our-story.

Gudgel, Mark. Interview with Keith Baker at Spring Mountain Vineyard. October 6, 2022.

————. Interview with Christopher Barefoot at Opus One. October 4, 2022.

————. Interview with Bob Biale at Robert Biale Winery. October 8, 2022.

————. Interview with Gustavo Brambila at Gustavo Wine Tasting Room. October 7, 2022.

————. Interview with Anna Brittain at Bistro Don Giovanni. October 7, 2022.

————. Interview with Paul Draper at Ridge. October 10, 2022.

————. Interview with Geoff Ellsworth at Vermeil Wines Tasting Room. October 5, 2022.

————. Interview with Mike Farmer, Mike Meneghelli and Lucas Farmer at Farmer residence. October 10, 2022.

————. Interview with Stacey Garrett at Chalone Vineyard. October 9, 2022.

————. Interview with Violet Grgich at Grgich Hills Estate. October 6, 2022.

————. Interview with Dennis Groth at Groth Vineyards and Winery. October 5, 2022.

————. Interview with Grace Hoffman at Freemark Abbey. October 7, 2022.

————. Interview with Kris Kraner at Mayacamas Vineyards and Winery. October 4, 2022.

————. Interview with Zelma Long via Zoom. November 2, 2022.

————. Interview with Sir Peter Michael via Zoom. September 20, 2022.

————. Interview with Vincent Morrow via Zoom. November 2, 2022.

————. Interview with Marcus Notaro via Zoom. November 3, 2022.

————. Interview with Bernard Portet at Clos du Val Winery. October 6, 2022.

————. Interview with Mark Rettig at Chateau Montelena. October 8, 2022.

————. Interview with Michael Silacci at Opus One. October 4, 2022.

————. Interview with geologist Angie Van Boening. September 22, 2022.

————. Interview with Nils Venge at Saddleback Cellars. October 5, 2022.

————. Interview with Emily Wines via Zoom. October 7, 2022.

Harris, Ren. "More Pelissa background." Personal communication with the author. November 28, 2022.

————. [No subject.] Personal communication with the author. November 28, 2022.

————. "Re: Your History in Napa." Personal communication with the author. November 18, 2022.

History.com Editors. "Mexican-American War." History.com. November 9, 2009. https://www.history.com/topics/mexican-american-war/mexican-american-war.

Larkmead Vineyards. "About." Accessed November 19, 2021. https://larkmead.com/pages/about/.

Library of Congress. "Mexican California." Accessed November 19, 2021. https://www.loc.gov/collections/california-first-person-narratives/articles-and-essays/early-california-history/mexican-california/.

Louis M. Martini. "Our History." Accessed January 22, 2022. https://www.louismartini.com/explore-martini/our-history.html.

Matasar, Ann B. *Women of Wine: The Rise of Women in the Global Wine Industry.* University of California Press, 2010.

Matthews, T. "The 1855 Bordeaux Classification." *Wine Spectator*, March 27, 2019. https://www.winespectator.com/articles/the-1855-bordeaux-classification-3491.

Napa County Historical Society. "Dr. Edward Turner Bale." July 23, 2015. https://napahistory.org/dr-edward-turner-bale/.

———. "Napa County History." Accessed November 17, 2021. https://napahistory.org/programs/local-history/napa-county-history/.

———. "Napa's First People." October 12, 2015. https://napahistory.org/napas-first-people/.

National Museum of the American Indian. "Treaty with the Potawatomi, 1836." Accessed November 19, 2021. https://americanindian.si.edu/nationtonation/treaty-with-the-potawatomi-1836.html.

National Park Service. "Bear Flag Revolt, June 1846." Accessed November 19, 2021. https://www.nps.gov/goga/learn/historyculture/bear-flag-revolt.htm.

National WWI Museum and Memorial. "Interactive Timeline." October 15, 2020. https://www.theworldwar.org/explore/interactive-wwi-timeline.

Native American Rights Fund. "Treaty of 1836." Accessed November 19, 2021. https://narf.org/nill//codes/grand_traverse/tre1836.pdf.

Oakville Winegrowers. "History." Accessed December 8, 2021. https://www.oakvillewinegrowers.com/h-i-s-t-o-r-y/.

O'Connor, Kelly. "RE: Rancho Question." Personal correspondence. January 7, 2022.

Parker, John. "1878 Painting of Elem Ceremony (By Jules Tavernier)." Wolf Creek Archaeology. Accessed January 20, 2022. http://www.wolfcreekarcheology.com/TavernierPainting.html.

Pryor, Alton. *The Mexican Land Grants of California*. Roseville, CA: Stagecoach, n.d.

Rancho Chimiles. "About." Accessed January 8, 2022. https://www.ranchochimiles.com/about.

Ridge Vineyards. "History." Accessed November 23, 2021. https://www.ridgewine.com/about/history/#:~:text=The%20history%20of%20Ridge%20Vineyards%20begins%20in%201885%2C,acres%20near%20the%20top%20of%20Monte%20Bello%20Ridge.

Serdar, Nina. "Winery History." Spring Mountain Vineyard. Accessed November 19, 2021. https://springmountainvineyard.com/a-wh.html.

SIMI Winery. "Learn About Our 145-Year-History." Accessed November 29, 2022. https://simiwinery.com/pages/about-us.

Spottswoode Winery. "The Estate." October 20, 2020. https://spottswoode.com/estate-overview/#history.

Stevenson, Robert Louis. *The Silverado Squatters*. Duke Classics, 2020.

Smith-Madrone. "Cook's Flat Reserve." Accessed November 29, 2022. https://smithmadrone.com/cooksflat/what.html.

Thompson, David. "Charles Krug Winery." The Napa Wine Project. Accessed November 13, 2021. https://www.napawineproject.com/charles-krug-winery/.

———. "Chateau Montelena." The Napa Wine Project. October 25, 2013. https://www.napawineproject.com/chateau-montelena/.

———. "Freemark Abbey Winery." The Napa Wine Project. Accessed November 16, 2021. https://www.napawineproject.com/freemark-abbey-winery/.

———. "Spring Mountain Vineyard." The Napa Wine Project. Accessed November 19, 2021. https://www.napawineproject.com/spring-mountain-vineyard/.

———. "Volker Eisele Family Estate." The Napa Wine Project. Accessed November 28, 2021. http://www.napawineproject.com/volker-eisele-family-estate/.

Van Boening, A. "Hey!" Personal communication with the author regarding geological formations of early California. September 22, 2022.

Weber, Lin. *Old Napa Valley: The History to 1900*. St. Helena, CA: Wine Ventures, 1998.

———. *Prohibition in the Napa Valley: Castles under Siege*. Charleston, SC: The History Press, 2013.

White, Kelli. "The Devastator: Phylloxera Vastatrix & The Remaking of the World of Wine." GuildSomm. December 30, 2017. https://www.guildsomm.com/public_content/features/articles/b/kelli-white/posts/phylloxera-vastatrix.

Chapter 2

Ballou, Jullianne. "From the Archives, Woman in Wine: Maryann Graf." UC Davis Library. March 8, 2019. https://library.ucdavis.edu/news/from-the-archives-woman-in-wine-maryann-graf/.

Beaulieu Vineyard. "100 Years of Quality." Accessed November 23, 2021. https://www.bvwines.com/en/100-years-of-quality-wine.html.

Chalone Vineyard. "Our History." Accessed November 23, 2021. https://www.chalonevineyard.com/About/Our-History.

Chateau Souverain. "Our Story." Accessed November 29, 2022. https://www.souverain.com/roots/story.

Clos du Val. "Une Histoire d'Amour." Accessed November 26, 2021. https://www.closduval.com/dig-deeper/une-histoire-damour/.

Ellsworth, F., and G. Ellsworth. "Robert "Bob" Ellsworth." Personal communication with the author. November 17, 2022.

Everett, L. "History of Heitz Martha's Vineyard." Personal communication with the author. November 10, 2022.

Freemark Abbey. "Jerry Luper and the University of Freemark Abbey." Accessed November 29, 2022. https://www.freemarkabbey.com/blog/freemark-abbey-legacy/jerry-luper-and-university-freemark-abbey.

Free Run Productions. "UNCORKED Season 1: The Judgement of Paris." January 19, 2017. YouTube video, 11:36. https://www.youtube.com/watch?v=IM1mel8DkBk.

Gump, James O. *Maestro: André Tchelistcheff and the Rebirth of Napa Valley*. Lincoln: University of Nebraska Press, 2021.

IMDb. "This Earth Is Mine." Accessed June 9, 2022. https://www.imdb.com/title/tt0053355/.

Los Angeles Times. "Peter Mondavi, Napa Valley Wine Pioneer, Dies at 101." February 22, 2016. https://www.latimes.com/local/obituaries/la-me-peter-mondavi-20160222-story.html.

Napa Valley Grapegrowers. "Mission & History." Accessed December 4, 2021. https://www.napagrowers.org/mission--history.html.

New York Times. "Rosa Mondavi, 86, Head of a Winery." Accessed December 2, 2021. https://www.nytimes.com/1976/07/06/archives/rosa-mondavi-86-head-of-a-winery-family-started-in-business-as.html.

———. "Transcript of the Toasts by Premier Chou and President Nixon." February 22, 1972. https://www.nytimes.com/1972/02/22/archives/transcript-of-the-toasts-by-premier-chou-and-president-nixon.html.

NilsVengeWine.com. "Nils' Timeline in Winemaking." Accessed December 2, 2021. https://nilsvengewine.com/about-nils.

Ridge Vineyards. "History." Accessed November 23, 2021. https://www.ridgewine.com/about/history/.

Robert Mondavi Winery. "The World of Robert Mondavi Winery." Accessed November 22, 2021. https://robertmondaviwinery.com/pages/about-us.

Romano, Aaron. "Peter Mondavi, Napa Valley Wine Pioneer, Dies at 101." *Wine Spectator*, February 22, 2016. https://www.winespectator.com/articles/peter-mondavi-napa-valley-wine-pioneer-dies-at-101-52774.

Schramsberg Vineyards. "History." September 20, 2016. https://www.schramsberg.com/about/history/.

Siler, Julia Flynn. *The House of Mondavi: The Rise and Fall of an American Wine Dynasty*. New York: Gotham Books, 2008.

Smith-Madrone. "Smith-Madrone Fact Sheet." Accessed November 27, 2021. https://www.smithmadrone.com/about/about.htm.

Sullivan, Charles L. *Napa Wine: A History from Mission Days to Present*. San Francisco: The Wine Appreciation Guild, 2008.

Swindell, Bill. "Wine Pioneer Mondavi Remembered as Strong-Minded Businessman." *Santa Rosa Press Democrat*. March 6, 2020. https://www.pressdemocrat.com/article/news/peter-mondavi-leader-of-charles-krug-winery-dies-at-101/.

Thompson, David. "Beaulieu Vineyard." The Napa Wine Project. Accessed November 23, 2021. https://www.napawineproject.com/beaulieu-vineyard/.

———. "Charles Krug Winery." The Napa Wine Project. Accessed November 13, 2021. https://www.napawineproject.com/charles-krug-winery/.

———. "Clos du Val." The Napa Wine Project. Accessed November 26, 2021. https://www.napawineproject.com/clos-du-val/.

———. "Freemark Abbey Winery." The Napa Wine Project. Accessed November 23, 2021. https://www.napawineproject.com/freemark-abbey-winery/.

———. "Heitz Cellar." The Napa Wine Project. Accessed November 23, 2021. https://www.napawineproject.com/heitz-wine-cellar/.

———. "Mayacamas Vineyards." The Napa Wine Project. Accessed November 24, 2021. https://www.napawineproject.com/mayacamas-vineyards/.

———. "Spring Mountain Vineyard." The Napa Wine Project. Accessed November 12, 2022. https://www.napawineproject.com/spring-mountain-vineyard/.

———. "Veedercrest Vineyards." The Napa Wine Project. October 25, 2013. https://www.napawineproject.com/veedercrest-vineyards/.

Trefethen Family Vineyards. "The Historic Winery." Accessed November 23, 2021. https://www.trefethen.com/learn/place/the-historic-winery/.

———. "The Trefethens." Accessed November 29, 2022. https://www.trefethen.com/learn/family/the-trefethens/.

Veedercrest Vineyards. "Veedercrest Is Born!" Accessed November 25, 2021. http://www.veedercrestvineyards.com/Veedercrest_Vineyards/Veedercrest_History.html.

Volker Eisele Family Estate. "Vineyard." Accessed November 26, 2021. https://volkereiselefamilyestate.com/place/.

Warren Winiarski. "Biography." Accessed November 25, 2021. https://www.warrenwiniarski.com/biography.

Waugh, Harry. *Diary of a Winetaster: Recent Tastings of French and California Wines.* New York: Quadrangle Books, 1974.

Wine Business. "California's First Woman Winemaker of the Modern Era, Mary Ann Graf, Passes." February 5, 2019. https://www.winebusiness.com/news/article/209274/.

WineCountry Staff. "The Judgment of Paris: 40 Facts for the 40th Anniversary." NapaValley.com. October 17, 2016. https://www.napavalley.com/blog/judgement-of-paris-facts/.

Chapter 3

Dickenson DePuy, J. *The Rest of the Story: The Paris Tasting.* N.p., 2016.

Gump, J.O. *Maestro: André Tchelistcheff and the Rebirth of Napa Valley.* Lincoln: University of Nebraska Press, 2021.

Hart, Steve. "French Judge Napa Wines Best." *Napa Register,* June 11, 1976.

Long, Zelma. "Re: The Judgment of Paris." Personal communication with the author. November 17, 2022.

Portet, Bernard. "Re: The Judgment of Paris." Personal communication with the author. November 17, 2022.

Rice, William. "Those Winning American Wines." *Washington Post,* June 13, 1976.

Spurrier, Steven. *Steven Spurrier: A Life in Wine.* Académie du Vin Library, 2021.

Taber, George. "Judgment of Paris." *TIME,* June 7, 1976.

———. *Judgment of Paris: California vs. France and the Historic 1976 Paris Tasting That Revolutionized Wine.* Scribner, 2006.

Wikipedia. "Judgment of Paris (Wine)." Accessed November 8, 2022. https://en.wikipedia.org/wiki/Judgment_of_Paris_(wine).

Chapter 4

Alcohol and Tobacco Tax and Trade Bureau. "American Viticultural Areas (AVAs)." Accessed January 28, 2022. https://www.ttb.gov/wine/american-viticultural-area-ava.

Berger, Dan. "Chalone, Rothschild Plan to Invest in Each Other." *Los Angeles Times*. February 10, 1989. https://www.latimes.com/archives/la-xpm-1989-02-10-fi-2300-story.html.

———. "It's Fun to Watch U.S. Beat Foreigners in Wine Olympics." *Oakland Tribune*, May 7, 1980.

Conaway, James. *Napa: The Story of an American Eden*. Boston: Mariner Books, 2002.

De Lacy, J. "The Great Wine Marathon: Gaston, They're Gaining on Us!" *International Herald Tribune*, 1979.

Dunn Vineyards. "The Story." Accessed November 15, 2022. https://www.dunnvineyards.com/story/.

Gaither, J'nai. "Phylloxera in Napa Valley: Then and Now." Wine Enthusiast, April 5, 2021. https://www.winemag.com/2021/04/05/napa-valley-phylloxera/.

Grgich Hills Estate. "40th Anniversary of the Judgment of Paris." January 19, 2016. YouTube video, 5:00. https://www.youtube.com/watch?v=7UQc7gm-sro.

Gudgel, Mark. Interview with Stu Smith via telephone. November 9, 2022.

IMDb. "Falcon Crest." Accessed November 12, 2022. https://www.imdb.com/title/tt0081858/.

je_admin. "Rules for Wine Labeling." Tracy Jong Law Firm. August 21, 2011. https://blog.tracyjonglawfirm.com/rules-for-wine-labeling.

Jones, Cory. "Want a Napa Valley Wedding? Read This First." Woman Getting Married. April 15, 2021. https://www.womangettingmarried.com/napa-valley-wedding-rules-laws.

MacDonald, G. "Re: Mondavi on To Kalon." Personal communication with the author. November 23, 2022.

MO Wine. "American Viticultural Areas." November 6, 2019. https://missouriwine.org/news/american-viticultural-areas.

Napa Register. "Trefethen's 1976 Chardonnay Takes Top Honors in Paris." October 13, 1979.

Parr, Devin. "An Introduction to Napa Valley AVAs." NapaValley.com. January 23, 2020. https://www.napavalley.com/blog/guide-to-napa-valley-avas/.

Peter Michael Winery. "About." June 25, 2019. Accessed November 15, 2022. https://petermichaelwinery.com/about/.

Prial, Frank J. "California Labels Outdo French in Blind Test." *New York Times*, June 9, 1976.

———. "Louis P. Martini, 79, Leader in Wine Industry." *New York Times*, September 23, 1998. https://www.nytimes.com/1998/09/23/us/louis-p-martini-79-leader-in-wine-industry.html.

Robards, T. "California Wines Again Win Prizes in a Blind Tasting Staged in France." *New York Times*, October 17, 1979.

Suckling, James. "Famous Napa Valley Winemaker Takes New Job in Portugal." *Wine Spectator*, August 20, 1999. https://www.winespectator.com/articles/ famous-napa-valley-winemaker-takes-new-job-in-portugal-20374.

Sullivan, Charles L. *Napa Wine: A History from Mission Days to Present*. San Francisco: Wine Appreciation Guild, 2008.

Trefethen Family Vineyards. "World Wine Olympics." August 13, 2020. https:// www.trefethen.com/journal/world-wine-olympics/.

Trefethen, J. "Re: Eshcol [*sic*] and my new book." Personal communication with the author. January 21. 2022.

————. "Trefethen's Early Success!" Personal communication with the author. February 11, 2022.

Wang, Ruoya. (2020). "Mitterrand's Monetary Policy: A Failure in the Economics Policy from 1981 to 1982." E3S Web of Conferences 214 (December 7, 2020). https://doi.org/10.1051/e3sconf/202021402010.

Wikipedia. "Judgment of Paris (Wine)." Accessed January 9, 2022. https:// en.wikipedia.org/wiki/Judgment_of_Paris_%28wine%29.

Wine Spectator. "Jack Davies, Napa's Outsider." June 15, 1998. https://www. winespectator.com/articles/jack-davies-napas-outsider-7737.

————. "Napa Valley Legend Joe Heitz Dies." December 18, 2000. https://www. winespectator.com/articles/napa-valley-legend-joe-heitz-dies-20882.

Winter, Mick. "Napa River Watershed Task Force." *Wine Business Monthly*, December 2000. https://www.winebusiness.com/ wbm/?go=getArticle&dataId=4903.

Chapter 5

Académie du Vin Library. "Wine Statistics." 2021, May 14, 2021. https:// academieduvinlibrary.com/wine-statistics.

Adams, Andrew. "Treasury to Buy Diageo Wine Brands for $552M." Wines Vines Analytics. November 2015. https://winesvinesanalytics.com/features/ article/160111/Treasury-to-Buy-Diageo-Wine-Brands-for-552M.

Barefoot. "Our History." Accessed October 30, 2022. https://www.barefootwine. com/our-story.html.

Bâtonnage. "Bâtonnage." Accessed October 29, 2022. https://www. batonnageforum.com/.

Belshaw, J.G. "Re: Identifying Fire Photo." Personal correspondence with the author. November 9, 2022.

Blazina, Carrie, and Drew DeSilver. "A Record Number of Women Are Serving in the 117th Congress." Pew Research Center. January 22, 2021. https://www. pewresearch.org/fact-tank/2021/01/15/a-record-number-of-women-are- serving-in-the-117th-congress/.

Booth, Edward. "Napa to Celebrate Dia de los Muertos." *Napa Valley Register*, October 31, 2022.

———. "Napa's Unhoused Residents Try to Survive Heat Wave." *Napa Valley Register*, September 9, 2022.

Bride Valley. "About Bride Valley Vineyard." November 11, 2021. https://bridevalleyvineyard.com/about-the-vineyard/.

Brittain, Anna. "History of Napa Green." Personal correspondence with the author. October 21, 2022.

———. "Re: Napa Green and the Judgment of Paris." Personal correspondence with the author. October 14, 2022.

Brown Estate. "Brown Estate." Accessed November 16, 2022. https://www.brownestate.com/#story.

Business Journal Staff. "Gallo Buys Napa Valley's Orin Swift Cellars." *North Bay Business Journal*, June 9, 2016. https://www.northbaybusinessjournal.com/article/industry-news/ej-gallo-winery-acquires-napa-valleys-orin-swift-cellars/.

California Department of Forestry and Fire Protection. "LNU Lightning Complex (Includes Hennessey, Gamble, 15-10, Spanish, Markley, 13-4, 11-16, Walbridge) Incident." Accessed October 20, 2022. https://www.fire.ca.gov/incidents/2020/8/17/lnu-lightning-complex-includes-hennessey-gamble-15-10-spanish-markley-13-4-11-16-walbridge/.

California State Parks, State of California. "Robert Louis Stevenson State Park." Accessed February 24, 2022. http://www.parks.ca.gov/?page_id=472.

Calkins, Victoria. *The Wappo People*. Santa Rosa, CA: Calkins, Ink, 1994.

Carl, Tim. "Beckstoffer, Cult Vineyards, and the Ascension of the Napa Valley." *Napa Valley Register*, September 22, 2022. https://napavalleyregister.com/wine/beckstoffer-cult-vineyards-and-the-ascension-of-the-napa-valley/article_c309dd80-3a01-11ed-86f7-7b3ae945690b.html.

Carson, L. Pierce. "Gallo Family Buys Out Louis M. Martini Holdings." *Napa Valley Register*, September 10, 2002. https://napavalleyregister.com/news/gallo-family-buys-out-louis-m-martini-holdings/article_7b6767ab-2c93-54cd-86e2-5c5dcff9c17a.html.

CIA Foodies. "Our Story: A History of the CIA." October 21, 2019. https://www.ciafoodies.com/cia-history/.

Clos du Val Napa Valley. "Carmel Greenberg Winemaker." Accessed October 29, 2022. https://www.closduval.com/wp-content/uploads/2021/10/Carmel-Greenberg-Bio.pdf.

Court of Master Sommeliers Americas. "Board of Directors." Accessed September 20, 2022. https://www.mastersommeliers.org/masters/board.

Donne, J., R.S. Hillyer and W. Blake. "A Lecture upon the Shadow." In *The Complete Poetry and Selected Prose of John Donne and the Complete Poetry of William Blake*. Modern Library, 1946.

Duarte, Jesse. "Napa Valley Harvest Report: Here Comes the Heat." *Napa Valley Register*, September 1, 2022. https://napavalleyregister.com/wine/napa-valley-harvest-report-here-comes-the-heat/article_65edf54a-295e-11ed-8ea9-930d737d25e5.html.

———. "Napa Valley Harvest Report: The Heat Is On." *Napa Valley Register*, September 8, 2022. https://napavalleyregister.com/wine/napa-valley-harvest-report-the-heat-is-on/article_cd9e99a4-2e33-11ed-8aed-837a48e64a14.html.

Eberling, Barry. "County Declares Tree Die-Off Emergency." *Napa Valley Register*, September 14, 2022.

———. "Napa Looks at Its Future Wine Country." *Napa Valley Register*, September 8, 2022. https://napavalleyregister.com/eedition/page-a1/page_7fcbaa91-cb0a-5a70-b8a1-d5100bc2a20a.html.

Eberling, G. "Electric Vine Buses Debut." *Napa Valley Register*, September 21, 2022.

Frank, Mitch. "Constellation Buys the Prisoner Wine Company for $285 Million." *Wine Spectator*, April 6, 2016. https://www.winespectator.com/articles/constellation-buys-the-prisoner-wine-company-for-285-million-52992.

"Extensions of Remarks." *Congressional Record* 146, no 23 (March 6, 2022). https://www.congress.gov/crec/2000/03/06/CREC-2000-03-06.pdf.

Franson, Paul. "French Influence in California/Pacific Northwest's Wine Business." Consulat Général de France à San Francisco. April 2018. https://sanfrancisco.consulfrance.org/french-influence-in-california-pacific-northwest-s-wine-business.

Frontline Wildlife Defense. "2020 California Lightning Complex Fires: Causes & Statistics." May 11, 2022. https://www.frontlinewildfire.com/wildfire-news-and-resources/2020-california-lightning-complex-fires/.

Gibb, Rebecca. "Chilean Wines Win 'Judgment of Los Angeles.'" Wine-Searcher. July 5, 2012. https://www.wine-searcher.com/m/2012/07/in-the-right-corner-chile-in-the-left-the-big-boys.

Goldman, Justin. "Good Judgment: Tasting Wine with Steven Spurrier." Hemispheres, March 5, 2021. https://www.hemispheresmag.com/inspiration/food-drink/tasting-wine-steven-spurrier-napa/.

Goldschmidt Vineyards. "Winemaker." November 12, 2019. https://goldschmidtvineyards.com/winemaker/.

Goode, Jamie. *Regenerative Viticulture*. n.p., 2022.

Gray, W. Blake. "Overturning the Judgment of Paris." Wine-Searcher. October 30, 2015. https://www.wine-searcher.com/m/2015/10/overturning-the-judgment-of-paris.

Grgich, M. "A Note about Bottle Shock." Napa Valley. February 4, 2016.

Gudgel, Mark. "History-Making Vino." *Dine Nebraska*, September 30, 2020. https://dinenebraska.com/2020/09/history-making-vino/.

———. Phone interview with Stu Smith. August 30, 2022.

———. Zoom interview with Sir Peter Michael. September 20, 2022.

———. Zoom interview with Victoria Coleman. November 15, 2022.

Gustavo Wine. "Our Winemaker." Accessed March 10, 2022. https://www.gustavowine.com/gustavo-brambila.

Harris, Ren. "Hello!" Personal communication with the author. November 2, 2022.

Heitz Cellar. "People." March 18, 2022. Accessed November 16, 2022. https://www.heitzcellar.com/people/.

Hill, E., A. Tiefenthäler, C. Triebert, D. Jordan, H. Willis and R. Stein. "How George Floyd Was Killed in Police Custody." *New York Times*, June 1, 2020. https://www.nytimes.com/2020/05/31/us/george-floyd-investigation.html.

Hinchliffe, Emma. "Female CEOS Run Just 4.8% of the Fortune Global 500." *Fortune*, August 3, 2022. https://fortune.com/2022/08/03/female-ceos-global-500-thyssenkrupp-martina-merz-cvs-karen-lynch/.

"H.Res.734—114th Congress (2015–2016): Recognizing and Honoring the Historical Significance of the 40th Anniversary of the Judgment of Paris and the Impact of the California Victory at the 1976 Paris Tasting on the World of Wine and the United States Wine Industry as a Whole." https://www.congress.gov/bill/114th-congress/house-resolution/734.

IMDb. *Bottle Shock*. Accessed October 21, 2022. https://www.imdb.com/title/tt0914797/.

Interview Area. "How Much Did Orin Swift Sell For?" Accessed October 30, 2022. https://www.interviewarea.com/frequently-asked-questions/how-much-did-orin-swift-sell-for.

Jones, Nicholas, Rachel Marks, Roberto Ramirez and Merarys Ríos-Vargas. "2020 Census Illuminates Racial and Ethnic Composition of the Country." United States Census Bureau. August 12, 2021. https://www.census.gov/library/stories/2021/08/improved-race-ethnicity-measures-reveal-united-states-population-much-more-multiracial.html.

Jones, Sam. "Napa's First Black Woman Winemaker, Victoria Coleman, Encourages Others to Pursue Viticulture." *Napa Valley Register*, March 10, 2022. https://napavalleyregister.com/wine/napa-s-first-black-woman-winemaker-victoria-coleman-encourages-others-to-pursue-viticulture/article_717863e1-3b7e-56a7-bd52-e3c5ce04160e.html.

Jordan, Michael T. "What You Should Know About Soaring Land Prices in Napa Valley." County Appraisals Inc. July 19, 2017. https://countyappraisals.net/2017/07/what-you-should-know-about-soaring-land-prices-in-napa-valley/.

Kamen, Robert. "JoP." Personal communication with the author. November 14, 2022.

———. [No subject.] Personal communication with the author. November 11, 2022.

———. "Re:." Personal communication with the author. November 11, 2022.

———. "Re: JoP." Personal communication with the author. November 15, 2022.

La Sirena Napa Valley. "Winemaker & Owner, Heidi Peterson Barrett." Accessed October 29, 2022. https://www.lasirenawine.com/Story/Winemaker.

Laube, James. "Iconic Napa Winery Stag's Leap Sold for $185 Million." *Wine Spectator*, July 30, 2007. https://www.winespectator.com/articles/iconic-napa-winery-stags-leap-sold-for-185-million-3679.

Laube, James, and Tim Fish. "Brother Timothy, a California Wine Pioneer, Dies at 94." *Wine Spectator*, December 1, 2004. https://www.winespectator.com/articles/brother-timothy-a-california-wine-pioneer-dies-at-94-2295.

Lewis, Kathleen. "Reaching out about CMS." Personal communication with the author, November 15, 2022.

———. "Re: Reaching out about CMS." Personal communication with the author, November 16, 2022.

Lobo Wines. "Victoria Coleman." January 2, 2022. https://www.lobowines.com/about/victoria-coleman/.

Lopez, G. "A Summer of Climate Disasters: Climate Change Has Made Extreme Weather Increasingly Normal." *New York Times*, September 7, 2022. https://www.nytimes.com/2022/09/07/briefing/climate-change-heat-waves-us-europe.html.

Los Angeles Times. "Kincade Fire Creates 200,000 Evacuees. 'Who's Going to Take Them In?'" October 29, 2019. https://www.latimes.com/california/story/2019-10-29/la-me-kincade-fire-evacuation-san-francisco.

Maisons Marques & Domaines. "People." Accessed October 29, 2022. https://mmdusa.net/portfolio/dominus-estate/people.

Marsteller, Daniel. "Exclusive: Foley Family Wines Buys Napa's Silverado Vineyards." Wine Spectator, June 21, 2022. https://www.winespectator.com/articles/foley-family-wines-buys-napas-silverado-vineyards.

Materra Wines. "Our Winemaking Team." Accessed October 29, 2022. https://www.materrawines.com/STORY/Winemaking-Team.

McGlauflin, Paige. "Meet the 6 Black CEOS on the Fortune 500." *Fortune*, May 25, 2022. https://fortune.com/2022/05/23/meet-6-black-ceos-fortune-500-first-black-founder-to-ever-make-list/.

Meeks, Alexandra. "A Record-Setting Heatwave Is Scorching the Western US." CNN. September 7, 2022. https://view.newsletters.cnn.com/messages/1662 546736163a89c85dc273d/raw?utm_term=1662546736163a89c85dc273d& utm_source=cnn_Five%2BThings%2Bfor%2BWednesday%2C%2BSeptem ber%2B7%2C%2B2022&utm_medium=email&bt_ee=AZ80E7Cov829MW qaEUQsVAYfS8nMv56eAq9MSQr1%2FglktgSECNGnKdJeUnrl%2BpDs& bt_ts=1662546736165.

me too. "History & Inception." July 16, 2020. https://metoomvmt.org/get-to-know-us/history-inception/.

Morrow, Vincent. "Re: Thank you so much." Personal communication with the author, November 4, 2022.

Moskin, Julia. "The Wine World's Most Elite Circle Has a Sexual Harassment Problem." *New York Times*, October 29, 2020. https://www.nytimes.com/2020/10/29/dining/drinks/court-of-master-sommeliers-sexual-harassment-wine.html.

Murphy, Linda. "Robert Mondavi Winery: What the Mondavis Did Next?" Decanter. May 30, 2007. https://www.decanter.com/features/robert-mondavi-winery-what-the-mondavis-did-next-247402/.

Myatt, Gary. *Designing and Painting Murals*. Ramsbury: Crowood Press, 2019.

———. "Introduction." September 6, 2021. https://garymyatt.com/.

———. "Re: Hello!" Personal communication with the author, June 26, 2022.

Napa Ag Preserve. "Ag Preserve Timeline." November 1, 2017. https://napaagpreserve.org/ag-preserve-timeline/.

Napa County California. "Conservation Regulations." Accessed October 30, 2022. https://www.countyofnapa.org/DocumentCenter/View/3374/Conservation-Regulations-Brochure-PDF.

Napa Green. "Sustainability Leadership Pillar: Social Equity, Justice & Inclusion." ThisIsCertifiedSustainable.Wine. Accessed October 19, 2022. https://thisiscertifiedsustainable.wine/social-equity/.

Napa Register. "Valley Is Flaming Nightmare." September 21, 1964.

Napa Valley Register. "Mayacamas to Open Tasting Room at First Street Napa Complex." Last updated March 27, 2019. https://napavalleyregister.com/mayacamas-to-open-tasting-room-at-first-street-napa-complex/article_c40e3785-159e-575f-8f08-f33f096cefe0.html.

Napa Valley Vintners. "History of the Napa Valley Vintners." Accessed October 19, 2022. https://napavintners.com/about/history.asp.

National Museum of American History. "American History (After Hours): The Judgement of Paris and American Wine." May 16, 2016. https://www.si.edu/object/yt_daUmHLeBrl0.

———. "Immigrant Winemaker." March 1, 2021. https://americanhistory.si.edu/food/wine-table/immigrant-winemaker.

———. "Judgment of Paris." March 1, 2021. https://americanhistory.si.edu/food/wine-table/judgment-paris.

———. "Napa Vintner." August 27, 2021. https://americanhistory.si.edu/food/wine-table/napa-vintner.

———. "Old Vines and New Blood." March 3, 2021. Accessed October 22, 2022. https://americanhistory.si.edu/food/wine-table/old-vines-and-new-blood.

New Hampshire Liquor and Wine Outlet. "Jean-Charles Boisset & Gina Gallo: The Wine World's Dynamic Duo." August 7, 2020. https://explore.liquorandwineoutlets.com/jean-charles-boisset-gina-gallo-the-wine-worlds-dynamic-duo/.

North Bay Business Journal. "Ohio Family Regains Full Ownership in Napa Valley Winery." September 11, 2017. https://www.northbaybusinessjournal.com/article/industry-news/ohio-family-regains-full-ownership-of-napas-mayacamas-vineyards/.

Paradigm Winery. "Winemaking." May 31, 2022. https://paradigmwinery.com/winemaking-2/.

Prchal, D. *Josephine Marlin Tychson: The First Woman Winemaker in California.* Napa County Historical Society, 1986.

Prial, Frank J. "Richard Graff, California Vintner, 60." *New York Times,* January 14, 1998. https://www.nytimes.com/1998/01/14/us/richard-graff-california-vintner-60.html.

Ridge Vineyards. "Winemaking Team." July 18, 2022. https://www.ridgewine.com/about/explore/winemaking-team/.

Rodgers, J., and G. Greschler. "Rain Brings Relief." *Napa Valley Register,* September 19, 2022.

San Francisco Recreation and Parks. "Coit Tower." Accessed October 25, 2022. https://sfrecpark.org/facilities/facility/details/Coit-Tower-290.

Schaeffer, Katherine. "Racial, Ethnic Diversity Increases Yet Again with the 117th Congress." Pew Research Center. January 28, 2021. https://www.pewresearch.org/fact-tank/2021/01/28/racial-ethnic-diversity-increases-yet-again-with-the-117th-congress/.

Smith, Rod. "Who Owns Napa? You Just Might Be Surprised." *Los Angeles Times*, June 27, 2001. https://www.latimes.com/archives/la-xpm-2001-jun-27-fo-15125-story.html.

Sogg, Daniel. "Chateau Montelena Sale Falls Through." *Wine Spectator*, November 5, 2008. https://www.winespectator.com/articles/chateau-montelena-sale-falls-through-4445#.

Stag's Leap Wine Cellars. "Our Story." Accessed October 18, 2022. https://www.stagsleapwinecellars.com/our-story/history.

Sweeney, Cynthia. "Boisset Purchases Calistoga Depot, Adding Another Napa Valley Location to the JCB brand." *Napa Valley Register*. Last updated March 24, 2022. https://napavalleyregister.com/news/local/boisset-purchases-calistoga-depot-adding-another-napa-valley-location-to-the-jcb-brand/article_8080a323-9a07-56e3-9bb2-400bf6a9d219.html

Thompson, David. "Charles Krug Winery." The Napa Wine Project. Accessed February 24, 2022. https://www.napawineproject.com/charles-krug-winery/.

———. "The CIA at Copia." The Napa Wine Project. Accessed March 9, 2022. https://www.napawineproject.com/the-cia-at-copia/.

———. "Gustavo Wine." The Napa Wine Project. Accessed March 10, 2022. https://www.napawineproject.com/gustavo-wine/.

———. "Mumm Napa." The Napa Wine Project. Accessed June 23, 2022. https://www.napawineproject.com/mumm-napa/.

Todorov, K. "Napa Valley's Spring Mountain Vineyard May Be Sold." Wine Business. Accessed November 29, 2022. https://www.winebusiness.com/news/article/264258.

Treasury Wine Estates. "Brands." Accessed October 18, 2022. https://www.tweglobal.com/brands.

Vinography (blog). "The Re-Judgment of Paris Results in California Landslide." April 30, 2020. https://www.vinography.com/2006/05/the_rejudgment_of_paris_result.

Welch, J. "The Great Fire." *Sonoma Democrat*, October 22, 1870.

The White House. "Kamala Harris." July 12, 2022. https://www.whitehouse.gov/administration/vice-president-harris/.

Wilde, Danielle. "Scholarships Aim at Wine Industry Diversity." *Napa Valley Register*, November 28, 2022. https://napavalleyregister.com/eedition/page-a1/page_b6624acb-38d4-5530-9cea-f3d3f025fc7d.html.

Willsher, Kim. "Hollywood Goes Nose to Nose over French Wine's Darkest Moment." *Guardian*, August 1, 2007. https://www.theguardian.com/world/2007/aug/01/usa.france.

Winter, Mick. "Who Owns Napa Valley's Vineyards?" *Wine Business Monthly*, May 2001. https://www.winebusiness.com/wbm/?go=getArticle&dataId=8217.

Wise, Jason. *Somm*. Forgotten Man Films, 2012.

———. *Somm 3*. Forgotten Man Films, 2018.

Women of the Vine & Spirits. "Home." October 6, 2022. https://www.womenofthevine.com/cpages/home.

Workman, Daniel. "Wine Exports by Country." Accessed May 5, 2022. https://www.worldstopexports.com/wine-exports-country/.

World Population Review. "Napa County, California Population 2022." Accessed February 23, 2022. https://worldpopulationreview.com/us-counties/ca/napa-county-population.

Worobiec, MaryAnne. "E. & J. Gallo Purchases Wine Brand Orin Swift Cellars." *Wine Spectator*, June 9, 2016. https://www.winespectator.com/articles/gallo-purchases-orin-swift-cellars.

———. "Iconic Napa Valley Winery Heitz Cellars Sold." *Wine Spectator*, April 18, 2018. https://www.winespectator.com/articles/napa-valley-winery-heitz-cellars-sold.

Yune, H. "Police: Napa Protest and March Against Police Violence Ended Peacefully After 8 Hours." *Napa Valley Register*, August 7, 2020. https://napavalleyregister.com/news/local/police-napa-protest-and-march-against-police-violence-ended-peacefully-after-8-hours/article_4f983eb0-1837-533d-9785-8cab88989cea.html.

Zimpfer, Zoe. "Re: Bernard and Carmel." Personal communication with the author, November 9, 2022.

INDEX

Michael, Sir Peter 78, 95, 116, 134, 135, 136, 137
Miravalle 35, 37, 49, 50, 77, 145, 157
Mondavi, Cesare 55, 56, 57, 58, 64, 116, 121, 155
Mondavi, Peter 55, 56, 64, 65, 67, 118, 134, 135, 136, 137, 155
Mondavi, Robert 54, 64, 65, 66, 76, 102, 105, 106, 114, 116, 118, 120, 122, 123, 125, 132, 138, 141, 155
Mondavi, Rosa 56, 58, 65, 121
Monte Bello 45, 83, 93, 111, 122, 153, 162
Montes 133
Morgan, Kathryn 152
Morrow, Vincent 151
Mount Saint Helena 22, 39, 54, 120, 142, 166
Mount Veeder 54, 145, 157
Mumm Napa 123
Myatt, Gary 134, 136, 137

N

Napa County 13, 29, 32, 121, 139, 142, 143, 144
Napa Green 140, 141, 179
Napa River 19, 20, 22, 24, 42, 72, 102, 120, 146, 166
Napa Valley 17, 19, 23, 24, 25, 28, 29, 31, 32, 35, 37, 39, 42, 43, 45, 46, 49, 50, 51, 54, 55, 56, 57, 58, 59, 61, 65, 66, 67, 68, 74, 76, 78, 82, 92, 95, 99, 100, 102, 103, 106, 107, 108, 114, 116, 117, 118, 120, 121, 124, 125, 127, 129, 132, 140, 141, 142, 143, 147, 148, 149, 151, 152, 153, 155, 157, 159, 162, 163, 166
Napa Valley Grape Growers Association 78
Napa Valley Vintners 102, 108, 140, 141, 152

Napa Valley Wine Technical Group 67
National Museum of American History 138
Nestlé 103
Niebaum, Gustave 37, 78
North American Plate 22
Notaro, Marcus 159
Novak, Mary 116
Nuns Fire, the 143

O

Oak Knoll 40
Oakville 37, 65, 78, 115, 123, 144, 153
Odyssey, The 99
Old Kraft Winery 37, 116
Oliver, Raymond 87, 135
Olympiad du Vin 107, 108
Onasatis 23
Opus One 114, 115, 141, 144, 153
Orin Swift Cellars 125
Ortman, Charles 77
Otsuka Pharmaceutical 116
Oxbow Market 127

P

Pacific 19, 22, 24, 25, 34, 42, 77, 102, 143
Pacific Ocean 19
Paradigm 153
Paris tasting. See Judgment of Paris
Parrott, John 35
Parrott, Tiburcio 35, 39, 135, 157
Paschich, Lee 66, 67
Patchett, John 32
Pelissa, Andrew 64, 103
Pelissa, Giuseppe 46
Pellett, Henry 34
Penfolds 133
Perrone, Osea 45, 47, 59
Petroski, Dan 148
Phillips, Jean 115, 153
Phinney, Dave 125

Veedercrest Vineyards 74
Venge, Nils 65, 72, 78, 115, 146
ViewCraft 140
Villaine, Aubert 87, 101, 134
Vineyard Hotel 134
Vrinat, Jean-Claude 87, 101, 134

W

Wappo 19, 22, 23, 24, 26, 50, 102,
    121, 142, 155
Weaver, Annie 34
Wine Hall of Fame 138
Wine Olympics. *See* Olympiad du Vin
Wines, Emily 150, 151, 152
Winiarski, Barbara 62, 63, 68, 76, 94,
    102, 125, 139
Winiarski, Warren 62, 65, 67, 68, 70,
    74, 94, 95, 100, 109, 114, 122,
    124, 129, 130, 138, 139, 153,
    159
Winroth, Jon 79
Wintun 19

Y

Yolo County 19
Yount, George 25, 26, 29, 31, 32, 34,
    50, 78, 116, 119, 123, 155
Yountville 15, 35, 39, 56, 64, 70, 119,
    148
Yugoslavia 54, 100, 103

ABOUT THE AUTHOR

Dr. Mark Gudgel is a father, husband, college professor, marathon runner and author. He is a regular contributor to *Edible Marin & Wine Country* and *Napa Valley Life*. Gudgel's previous book, *Think Higher Feel Deeper*, focuses on his career in Holocaust education. Gudgel is currently working on a book about the Court of Master Sommeliers and a travel guide to the Napa Valley. He is the president of the board of the vinNEBRASKA Foundation, which helps raise money for charities. Gudgel lives in Omaha with his wife, Sonja, and their children, Zooey and Titus.